WASHOE COUNTY LIBRARY

3 1235 02961 9413

FEB 1 0 2005

D0762092

RENO
DOWNTOWN
BRANCH

SEATURTLES

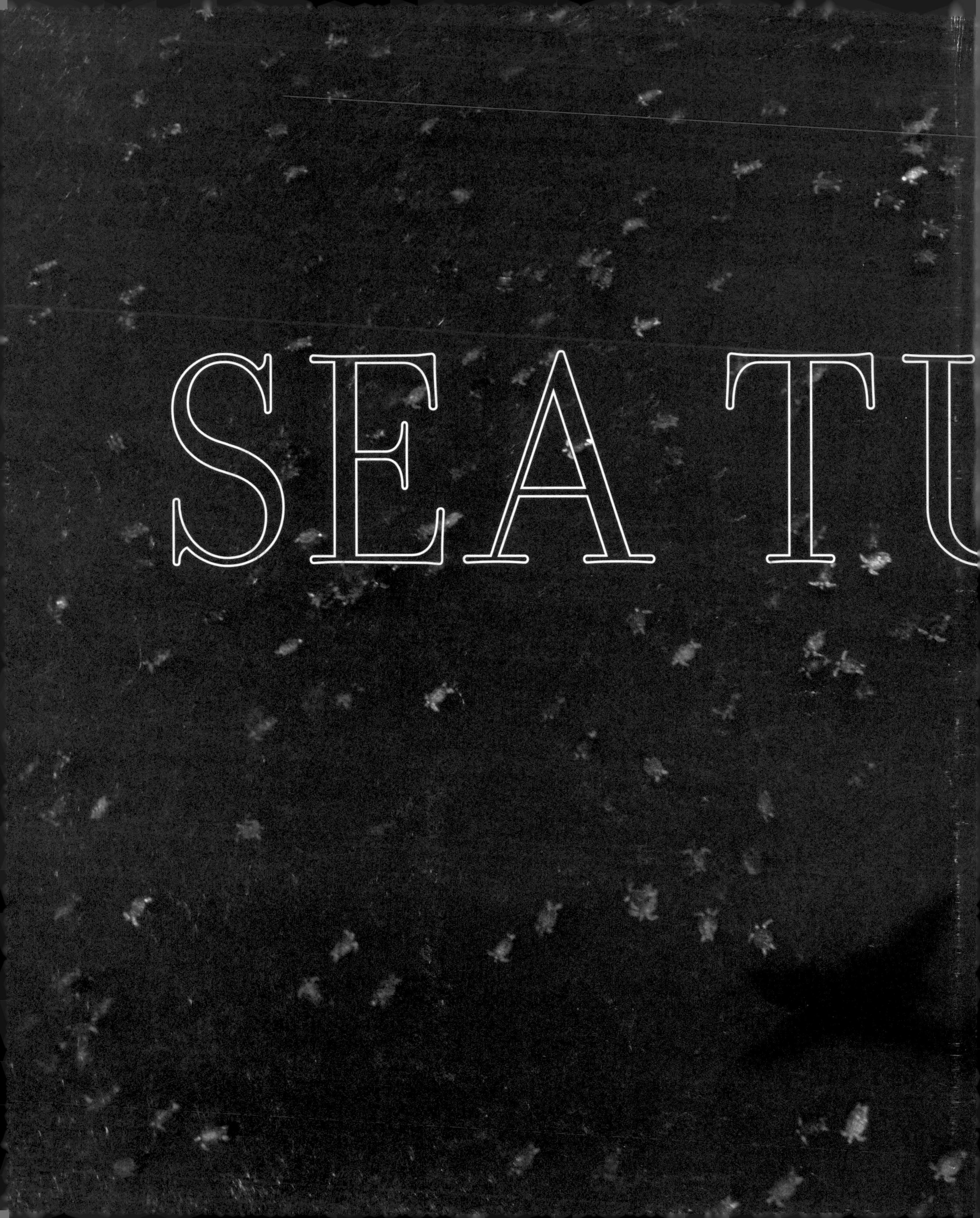
SEA TU

URTLES

A COMPLETE GUIDE TO THEIR BIOLOGY, BEHAVIOR, AND CONSERVATION

JAMES R. SPOTILA

The Johns Hopkins University Press

Baltimore and London

© 2004 The Johns Hopkins University Press and Oakwood Arts
All rights reserved. Published 2004
Printed in China on acid-free paper
9 8 7 6 5 4 3 2 1

The Johns Hopkins University Press
2715 North Charles Street
Baltimore, Maryland 21218-4363
www.press.jhu.edu

Library of Congress Cataloging-in-Publication Data
Spotila, James R., 1944-
    Sea turtles : a complete guide to their biology, behavior, and
conservation / James R. Spotila.
        p. cm.
    Includes bibliographical references and index (p.   ).
    ISBN 0-8018-8007-6 (hardcover : alk. paper)
    1. Sea turtles. I. Title.
QL666.C536S69 2004
597.92'8—dc22   2004008935

A catalog record for this book is available from the British Library.

Photography and illustration credits may be found on page 227,
which is an extension of this copyright page.

Title page photograph: Thousands of green turtles amassing
near Raine Island on the Great Barrier Reef to lay their eggs in
November and December.

To THE NEXT GENERATION, into whose hands we give the sea turtles: my children Jennifer and James and my nephews Alex Spotila, Maxwell Dittrich, and Marshall Dittrich.

CONTENTS

Pivoting with its left flipper, this hawksbill performs a sharp turn. Hawksbills have been heavily harvested for their beautiful shells, while their important role in maintaining coral ecosystems has gone underappreciated.

Green turtle with remoras (also called sharksuckers) attached, swimming along the reef at Sipadan Island off Sabah, Malaysia. The remoras are a specialized fish that attach with a sucker-like fin to large fish and sea turtles. Other than creating a little extra weight and drag, the remoras don't harm the turtles as far as we know.

PREFACE

THIS BOOK EVOLVED OVER A NUMBER OF YEARS, INSPIRED BY DISCUS-
sions with Victor Hutchison, but was delayed while I served in the Clin-
ton Administration as Chief Environmental Scientist for the Department
of the Army. For me it was a new world, complete with an office in the
Pentagon and the responsibility of trying to help a quite willing Army minimize
its negative impact on the environment. My boss, Ray Fatz, the Deputy Assistant
Secretary for the Army for Installations and Environment, took to calling me Dr.
Turtle in those days. Coming back to academia I was able to devote more time to
studying and conserving sea turtles again, and it quickly became obvious to me that
leatherback turtles were in very serious trouble indeed. Frank Paladino and I soon
formed a non-profit organization called The Leatherback Trust to raise funds and
focus our energies on conservation of the leatherback turtle. Once our new website,
www.leatherback.org, was created we began to get inquiries from people of all ages
about leatherbacks and other sea turtles: How many are there? When do they lay
their eggs? Where do they live? How deep do they dive? What do they eat? Questions
came in from third graders and engineers, high school students and grandmothers.
It seemed that there was a real need for an informative book on sea turtles.

About that time two critical things happened. First, my friends and former staff
members in the Pentagon narrowly survived the terrorist plane crash there on September 11, 2001. That was a wake-up call: Do the important things while you can
because you do not know how much time you have left. Second, I was fortunate
to receive encouragement from my editor, Vincent Burke. He assured me we really
could produce a book that would be not only informative but also interesting to read
and visually exciting. Now that it is completed I hope all of those who have been ask-
ing questions will find their answers within these pages.

This book was not one produced in isolation. It was a work of cooperation from
beginning to end. There are a number of people to whom I owe thanks. Without
the support of John and Renee Grisham's Oakwood Arts & Sciences Trust, this book
could not have been produced. Frank and Merry Thomasson diligently oversaw all
aspects of its design and production and manufacture. Mark Gatlin offered many
helpful comments as he copyedited the book. David Griffin created the book's beau-

tiful design. Val Kells and Dawn Witherington drew the lovely illustrations. Finally, I am indebted to a number of scientists who have reviewed chapters and provided information, particularly Pamela Plotkin, Karen Bjorndal, Anne Meylan, J. Whitfield Gibbons, Jeanne Mortimer, Patrick Burchfield, Charles Caillouet, Alan Bolten, Colin Limpus, Anders Rhodin, Jeffrey Seminoff, Jack Frazier, John Parmenter, Maria Marcovaldi, Kartik Shanker, Nicholas Pilcher, and anonymous reviewers.

I thank all of those who have made it possible for me to study, come to know, and to write this book about sea turtles. My mother, Mary, and father, John, inspired me to love the world and to work to make it better. My stepmother, Veronica, taught me that love has no bounds. My wife, Laurie, has been my source of love and strength, my constant companion, and she has made life worth living for over 37 years. She also provided support in the editorial process for the book. My colleagues and students have taught me more than I could ever have taught them. Frank Paladino has inspired me to work harder for the turtles. I could not have accomplished anything without the help of Ed Standora, Bob Foley, Steve Morreale, Georgita Ruiz, Mike O'Connor, Dave Penick, Alison Leslie, Tony Steyermark, Richard Reina, Robert George, David Rostal, Paul Sotherland, Janice Grumbles, Al List, Walter Bien, Hal Avery, Bruce Kingsbury, Christopher Binckley, Philip Mayor, Caitlin Curtis, Jennifer Crim, Barbara Bell, Dana Drake, David Reynolds, Karen Rankin-Baransky, Kris Williams, Bryan Wallace, Susana Clusella, Vince Saba, Maria Santidrian, and Jim Hansen. In addition, I thank the hundreds of students and volunteers who have dedicated themselves to the leatherback project in Costa Rica. Mario Boza, Maria Teresa Koberg, Clara Padilla, Enrique and Yanira Chacon, and many friends in Costa Rica have been an inspiration and have provided immeasurable assistance in my conservation work there. I am honored to be their colleague and friend. ❧

Royalties earned from the sale of this book will be donated to the Leatherback Trust.

SEATURTLES

CHAPTER ONE

INTRODUCTION
The Seven Swimmers

A LARGE, BLACK SHAPE DRIFTED CLOSER AND CLOSER, COMING TOWARD us through the surf. It was dark and rainy that March night in Tortuguero, Costa Rica. My friend Frank Paladino, a professor and one of my former students, had been walking the black sand beaches with me for hours looking for leatherback turtles. Soaked to our skins, we had flopped down against a large tree trunk to rest. We both spotted the log-sized shape at the same time but said nothing because we could barely see through the rain. Was it one of the many floating logs in those waters or was it alive, a creature from the depths? After our long, wet trek maybe our eyes were betraying us in the dark. The next crash of surf answered our questions: This was no log but rather an eight-foot-long sea turtle, very much alive.

As the great reptile approached we made out her huge head, which was the size of a watermelon. We spoke softly, confirming our suspicions. The surf glowed with the phosphorescent light of phytoplankton, the one-celled algae that form the basis of all life in the sea, as they cascaded over her shell. "It's a leatherback," we kept repeating as she headed right for us.

It was 1989, and Frank and I were there to study the physiology of the leatherback, the world's largest reptile. We often talked about leatherbacks as "modern dinosaurs" but did not realize then that the species was perilously close to joining its relatives in the oblivion of extinction.

A leatherback turtle returns to the sea at dawn after an arduous two hours on the beach laying her eggs at Playa Grande, Costa Rica.

The leatherback emerged from the water and lumbered up the beach with an effort that was exhausting to watch. She lifted her 800-plus pounds (360 kg) on her foreflippers and pushed forward with her rear flippers. We laid still against our tree trunk as she passed close by, oblivious to our presence. I motioned to Frank for us to move once she got a bit farther up the beach. Then like commandos we crouched low and followed her. She stopped soon after reaching the first sprigs of beach grass. We sat down on the sand, several yards behind her, and watched as she dug a shallow pit with her front flippers, then a deeper nest with her hind flippers. Her movements had an unmistakable rhythm—right, left, dig, throw—like some nineteenth-century machine. Sand sprayed gently to each side and two piles began to grow. Near our end of the pit her hind flippers slowly dug a hole about three feet deep. Then the sand stopped flying. We edged closer and watched as her soft, creamy-white eggs, each about the size of a billiard ball, dropped into the hole. We tried to count them, guessing at times, and estimated her clutch to be about 80 eggs.

After the last egg fell she covered them with sand and soon began to make her way back to the ocean. Our entertainment finished, Frank and I became scientists again. We caught up to her and quickly, while she moved, duct-taped on our special "leatherback mask," covering her nose and mouth. Part of our research involved measuring the metabolism of nesting leatherbacks, so we needed samples of the air she exhaled. What she thought of these two bipedal "naked apes" desperately trying to bag her breath I could only imagine.

Frank indicated that he had the data he needed so we unmasked her. She waddled back into the ocean just as the sun came up. Sitting back down on the beach I think we both felt the same sense of awe. We had collected our data as scientists, but as humans we had experienced the greater meaning of the moment.

Watching a sea turtle rise from the waves, lay its eggs, and then disappear into the churning ocean cannot help but tug at your heart. In the sand beneath us lay her offspring, unlikely to ever meet the mother that traveled hundreds, perhaps thousands of miles to that beach to select the one spot she thought safe from predators. There is something so primeval yet so basically sympathetic about these creatures that anyone who witnesses such an event cannot help but worry about their future.

As laborious as it is for sea turtles on land, there are few sights more majestic than a sea turtle gliding through a kelp forest or past a coral reef, propelled by seemingly effortless strokes of the foreflippers. The seven species alive today are ancient reptiles—living dinosaurs, if you will—swimming through our oceans just as they did one hundred million years ago. Sea turtles survived the mass extinction that wiped out the dinosaurs, yet they are not indestructible. Indeed, during the twentieth century we came close to losing the species known as the Kemp's ridley. These, among the smallest of the sea turtles, nest primarily on a single beach on Mexico's Gulf Coast. In the late 1940s at least 40,000 females were counted in a single day, but by the 1980s that number had dwindled to 300. Twenty years of concerted conservation efforts have stopped that decline. Now there are more than 5,000 adult females in the Kemp's ridley population.

The continuing story of the Kemp's ridley is encouraging, but trouble brews for other species. The nomadic leatherback, weighing as much as 2,000 pounds (907 kg), is so named because of the rubbery skin that covers its top shell. It seems to dine almost exclusively on jellyfish, somehow finding enough nutrition in this limpid prey to grow twice as large as the next largest species, the green turtle. Leatherbacks lay their eggs on five continents, but in the Pacific Ocean their numbers have declined from 90,000 nesting females in 1980 to fewer than 5,000 today. Rapacious beach development near nesting areas and deadly industrial fishing methods threaten to eliminate leatherbacks from Earth's largest ocean. Just when we have reason to hope Kemp's ridleys are on the rise, it appears that leatherbacks are on the brink of extinction.

This is the paradox of sea turtles. People everywhere hold them in high esteem, as symbols of grace, fertility, wanderlust, and long life. We work hard to save them and yet we are the murderous instruments of their doom. This book is my attempt to put into words what a lifetime of studying turtles has taught me. It is a complex story about a complex group of animals that have suffered stunning declines in the twentieth century. I am a scientist who became a conservationist because I could not turn my back on the preventable disasters that I saw looming for sea turtles. I also could not fathom a world without them. These are fascinating creatures, and I am wagering that the more you know about them, the more you will want to make sure that they are here for generations of us yet unborn.

The seven species of sea turtles are descendants of a small species that wandered out of freshwater marshes and entered the sea. Each of the species is unique, some in startling ways. The smallest, the Kemp's ridley, is named after Richard Kemp, one of the first people to study the species. The 'ridley' part of the name is a bit of a mystery. While the Kemp's ridley nests almost exclusively on a single beach in Mexico, the closely related and similarly-sized olive ridley nests on beaches in the Atlantic, Pacific, and Indian Oceans. Ridleys display a nesting phenomenon known as an *arribada*, or mass nesting. During an *arribada* as many as 100,000 turtles emerge from the surf to nest.

Green sea turtles are probably the best-known species because snorkelers and scuba divers can observe them in the shallow waters of the Caribbean, usually feeding on what is commonly called turtle grass. They nest throughout the tropics and subtropics, growing to several hundred pounds. We don't know how long they live, but we believe that they outlive humans and in some areas may not lay their first eggs until they are 35 years old or more. Green turtles often nest in Florida, as does the loggerhead. Like the ridleys, loggerheads are carnivorous, preferring a diet of shellfish and crabs. They are the most common sea turtle in the United States and are recognized by their tendency to ferry the ocean's biota. Epibionts, organisms that grow on other organisms, love loggerhead shells, which become little moving

Leatherback Turtle

Dermochelys coriacea

Shell length: 52-70 inches (132-178 cm)

Mass: 550-2000 lbs (250-907 kg)

Color: carapace is black with white splotches and plastron is white with black splotches

Features: The largest sea turtles, leatherbacks are covered with a continuous layer of thin rubbery skin instead of hard shell. The carapace is raised into a series of seven longitudinal ridges. Males reach a shell length of about 6.5 ft (198cm) and weigh up to 2000 lbs (907 kg). Leatherbacks eat jellyfish and range over all the oceans except the Arctic and Antarctic. They dive as deep as a whale in search of food.

Kemp's Ridley Turtle

Lepidochelys kempii

Shell length: 24-30 inches (61-76 cm)

Mass: 80-100 lbs (36-45 kg)

Color: olive green carapace and yellowish-brown plastron

Features: Slightly larger and heavier, lighter in color and with a lower and wider carapace than the olive ridley. Its head is large with powerful jaws. It ranges over the Caribbean Sea and the North Atlantic as far north as Massachusetts. Despite recent successes in conservation it remains the rarest and most endangered sea turtle. Kemp's ridleys feed along coastal shallows in depths of 150 feet (46 m) or less and eat crabs, clams, and snails. They return to the beach every year or two and lay their eggs before returning to wander the seas. Kemp's ridleys nest in an *arribada* at Rancho Nuevo on the Gulf coast of Mexico.

Olive Ridley Turtle

Lepidochelys olivacea

Shell length: 22-30 inches
   (55-76 cm)

Mass: 80-95 lbs (36-43 kg)

Color: olive green shell (carapace) above and light greenish-yellow shell (plastron) below

Features: The most abundant sea turtle. It ranges over much of the tropical Pacific, Indian and Atlantic Oceans. Its head is large with powerful jaws. Olive ridleys spend their adult lives swimming along drift lines, where debris and masses of floating seaweed (algae) are concentrated, and feeding on the animals available there. Olive ridleys eat a varied diet of crabs, jellyfish, clams, snails and some algae. Females come ashore in large groups called *arribadas*, meaning arrivals in Spanish.

Green Turtle

Chelonia mydas

Shell length: 32-48 inches
(80-122 cm)

Mass: 144-450 lbs
(65-204 kg)

Color: light or dark brown carapace sometimes shaded with olive, with bold streaks or blotches of brown, and a yellowish-white plastron; "black turtle" is considered a subspecies (*Chelonia mydas agassizii*) and has a dark gray to black carapace and a yellowish-white plastron.

Features: Green turtles have small rounded heads and a smooth carapace. These herbivores range throughout the tropical and subtropical Atlantic, Pacific and Indian Oceans where they eat sea grasses and rooted algae. Their name comes from the greenish color of their fat, which is due to their diet of sea grasses. Scientists have argued as to whether the "black turtle" is really a separate species, but recent DNA data and studies of skull morphology strongly indicate that it is too similar to other green turtles to be granted such status.

Hawksbill Turtle

Eretmochelys imbricata

Shell length: 30-35 inches (75-88 cm)

Mass: 95-165 lbs (43-75 kg)

Color: carapace is dark amber with radiating streaks of brown or black and plastron is whitish-yellow.

Features: Head is narrow with a strongly hooked beak that gives it its name. The carapace has thick overlapping scales (scutes) and is serrated along the posterior edge. Hawksbills live along coral reefs throughout the tropical oceans and eat sponges. They nest on tropical beaches.

Loggerhead Turtle

Caretta caretta

Shell length: 34-49 inches (85-124 cm)

Mass: 176-440 lbs (80-200 kg)

Color: carapace is reddish-brown and plastron is dull brown to yellowish

Features: Head is very large with strong crushing jaws. Loggerheads feed throughout the water column in shallow waters (less than 200 feet (61 m) deep) and eat horseshoe crabs (*Limulus*), crabs, mollusks (clams and sea snails), sea pens and assorted invertebrates. They feed in estuaries, along the continental shelf and in the open (pelagic) ocean where they eat crabs associated with drift lines and masses of sea weeds. Loggerheads nest on beaches of the subtropical oceans, along the Mediterranean shore, and on the beaches from southern Florida to South Carolina.

Flatback Turtle

Natator depressus

Shell length: 30-39 inches (75-99 cm)

Mass: 154-198 lbs (70-90 kg)

Color: carapace is olive-gray with pale brown/yellow tones on margins and plastron is cream colored

Features: Characterized by its flattened shell this is the least known of the sea turtles because its range is limited to the tropical waters of Australia and most of its nesting beaches are remote and undeveloped. It is distantly related to green turtles and loggerheads but is genetically quite distinct. Flatbacks feed in shallow, inshore waters on soft bottoms throughout northern Australia and appear to avoid reefs. They feed on jellyfish, sea pens and other soft-bodied bottom dwelling invertebrates.

Archie Carr: The Father of Sea Turtle Biology and Conservation

Archie Carr was ahead of his time. He not only studied the biology of animals, but also saw the need to conserve them. He was a keen observer of the natural world and used his literary skills to tell the world what he had learned. Almost single-handedly he brought sea turtles from the edge of extinction to the edge of hope.

Archie provided a model for others to follow and a legacy of followers to carry on his work. Most sea turtle biologists trace their roots, either directly or indirectly, to Archie Carr. The older ones were his students or worked with him during their careers. The younger ones studied or worked with Archie's academic offspring. Now the world is filled with Archie's academic grandchildren and great-grandchildren.

Archie was born in 1909 in Mobile, Alabama, and died at his home at Wewa Pond near Micanopy, Florida, in 1987. He began college as an English major but switched to science and earned his M.S. and Ph.D. degrees at the University of Florida. His books *The Windward Road* and *So Excellent a Fishe* startled the world and still fascinate readers today with their blend of science, folklore and humor. They provided the philosophical basis for a generation of sea turtle conservationists and led to the formation of the Caribbean Conservation Corporation.

In 1954 Archie Carr started the green turtle program at Tortuguero, Costa Rica. That program is the longest continuous population study of any vertebrate. Years later he participated in the first experiments using radio telemetry and promoted other new technology to study and conserve sea turtles.

Archie was always pushing the envelope of the possible to give sea turtles a break in their increasingly difficult world. He pushed to try to reestablish green turtle nesting colonies around the Caribbean, to foster the Kemp's ridley restoration project at Rancho Nuevo, Mexico, to encourage the first studies on temperature dependent sex determination in sea turtles and to focus efforts on the biology of leatherback turtles. He also accurately predicted where biologists would find juvenile sea turtles during their "lost years."

Sea turtles survive today in large part because of Archie Carr. He was one with his organism and also one with people. He was as comfortable with the wealthy and learned as he was with Carib, Tico, Miskito Indian or African fishermen. He publicized the plight of sea turtles but shunned publicity and the public eye. He disappeared around the back of the old Casa Verde at Tortuguero upon the approach of strangers, but took time from his busy day to talk to a ten-year-old boy who was fascinated by the little sea turtle swimming in the Carr laboratory. He was a combination of good heart and powerful intellect. His lasting hope was that the conscience of humanity would eventually save the wild things from obliteration.

islands covered with algae, barnacles, clams, small skeleton shrimp, and a host of other species.

Hawksbill turtles are unique among the sea turtles for feeding almost exclusively on sponges. They are the source of all tortoise shell jewelry, brush handles, and other such items because of their beautiful, overlapping shell plates. Although trade in such items is now illegal in most places, it was once so common that the hawksbill population has been greatly reduced in the 250 years since colonial times.

Another species of hard-shelled sea turtle can be found only in the waters around and on the beaches of Australia. Appropriately named the flatback turtle, having a soft slope to its shell, it feeds on soft-bodied invertebrates and nests in northern Australia.

Finally there is the leatherback, a very different kind of sea turtle. Its shell is pelleted with bone but consists mostly of cartilage and taut membranes that give it a rubbery feel. And these turtles are huge, often measuring eight feet from nose to tail. Nomadic wanderers, they swim far out to sea in search of jellyfish and often dive thousands of feet (700 meters) to feed or to cool off in tropical seas. They seem to be immune to the stingers of their prey and are the fastest growing of the seven species.

Ranging from hatchlings a few inches long to mature behemoths weighing thousands of pounds, the seven sea turtle species swim from the northern and southern reaches of the Atlantic and Pacific Oceans to the tropics and into the Mediterranean Sea. They go ashore to lay their eggs on Greek resort beaches and on isolated stretches of India's shore. They eat a wide-ranging diet, from sea grasses to jellyfish, algae, crabs, clams, and sponges. Yet all sea turtles are in danger of extinction and many populations will likely disappear within our lifetimes. We are loving our beaches to death, eliminating sea turtle nesting areas. And as we struggle to feed a hungry world, they die needlessly in industrial fishing operations. One by one we are removing them from the ocean.

Extinction, however, is not inevitable. Coexistence can be easily achieved with just a reasonable accommodation on the part of humanity. Progress is being made and there is hope, but our generation must take the critical actions necessary to save these extraordinary animals. Sea turtles are like the canary in the mineshaft: As they go, so go the oceans. It is my hope that we will all look back a decade or two hence and sigh with relief because the world's sea turtle populations are growing and on their way to returning to what they were a century ago.

There are heroes out there on the beaches and at sea who are devoting their lives to saving these animals for our grandchildren. Young and old, they are scientists, conservationists, government officials, and volunteers. They are moving nests in danger from high tides, guarding hatchlings on their crawl to the sea, working on board fishing boats to reduce turtle deaths in the fishing industry, and facing armed poachers and refusing them access to nesting turtles. The fate of sea turtles is in their hands but also in ours. There are many things that we can do, even without visiting the ocean. Time is short, the task is large, and the situation is complicated. ⌒

CHAPTER TWO

LIFE CYCLES

From Sand to Sea

WITH A SOFT PLUNK, PLUNK, PLUNK, THREE GOLF-BALL-SIZED EGGS dropped near the tail of a green turtle and fell to the bottom of the hole she had dug in the sand. The female turtle exhaled loudly and gasped in a lungful of air as she prepared to drop more eggs. She was calm, in what seemed like a trance, and oblivious to the scientists lying on the black sand beach behind her. We picked up one of her rear flippers and peered into the private world of her birth experience. We watched her body tense up as she squeezed two more eggs into the nest chamber. A clear, watery fluid covered the eggs and dripped out of her onto the growing clutch. This fluid lubricated the eggs as they passed out of the mother and kept them moist until she covered up the clutch with damp sand. For sea turtles, this is how it all begins.

Witnessing this moment of new life is an unforgettable experience. It is a laborious and difficult task, particularly for an animal that is at home only in the sea. Mother turtles crawl out of the ocean and dig nests for their eggs without being able to see what their rear flippers are doing. They strain to lay 50 to 100 eggs or more. Finally, after covering the eggs with sand, they drag their heavy bodies back to the sea, having spent two exhausting hours on land.

Often I feel compelled to guard over the nest, watching it for a while after the mother departs to protect the eggs from raccoons, coatis (South and Central American relatives of the raccoon), and coyotes. Perhaps it's a token effort, but it's one that is hard to resist.

Inside the Egg

All sea turtles lay eggs, a trait they share with other turtles and most reptiles. As such, they are oviparous (from the Latin *ovum parere*, or egg producer). Some reptiles retain their eggs inside their bodies and deliver live offspring into the world. But so far no sea turtle, or any other type of turtle for that matter, has evolved that trait. Perhaps the reason is that building nests has worked quite well for millions of years.

Deep inside the egg lies the embryo, the predecessor of the hatchling turtle. It is formed by joining one sperm from a male turtle with one ovum from a female. Adjacent to the embryo lie a sac of amniotic fluid, a yolk sac, and the allantois, a mem-

Previous pages: Loggerhead turtle hatchling heading to the ocean on Little Saint Simon's Island, Georgia. Emerging from the nest in the daytime is dangerous for a sea turtle because seabirds are effective visual predators.

Opposite: Sea turtles drop their flexible shelled eggs into the nest chamber. The eggs emerge from the cloacal opening in ones, twos, and threes accompanied by watery secretions that help to prevent bacterial and fungal infections.

branous sac that develops from the rear portion of the embryo's gut. Each fulfills a specific need for the embryo as it develops. Amniotic fluid provides a stable aquatic environment and absorbs shock waves, the yolk provides nourishment, and the allantois allows for gas exchange between the embryo and the air within the nest, as well as for removing waste. Blood vessels in the lining of the yolk sac transport fats, sugars, starches, and proteins to the embryo. Ammonia, urea, and other wastes pass out of the embryo through a passage that connects the rear portion of the gut to the allantois. As the embryo gets larger and produces more waste, the allantois grows. Balance is achieved within the egg because as the allantois grows the yolk sac shrinks.

Leatherback turtle eggs from a clutch collected in Irian Jaya, Indonesia, as part of a conservation project to protect them from poachers. The smaller "false eggs," or SAGs, contain albumen but no embryo.

Surrounding the embryo, allantois, yolk sac, and amnion is a membrane called the chorion, which is filled with blood vessels. Beyond the chorion lies the albumen (egg white), a watery medium that allows gasses to exchange through the shell and provides for the developing embryo's water needs. Sea turtle eggshells are not hard like bird eggs; they have the feel of leather or parchment. The shell is thin and porous enough to allow gases to pass through with relative ease—the embryo actually breathes through the shell.

Embryo metabolism is fueled by oxygen, which is combined with sugar from the yolk and eventually absorbed by the hemoglobin in the red blood coursing through the blood vessels in the chorion. In the process, carbon dioxide and water are produced. The carbon dioxide is released from the blood cells and passes out through the shell into the nest and ultimately up through the sand to the air above. A water molecule is so small—smaller even than that of oxygen and carbon dioxide—that it passes through the shell quite easily. Over the course of about two months, gasses entering the shell are balanced with the volume leaving the shell, allowing the turtle to fill the eggshell without splitting it open.

How fast an embryo becomes a hatchling sea turtle is determined largely by temperature. If the sand is cool (77°F/25°C) because the nest is in the shade or there have been prolonged periods of rain, incubation can take 65 to 70 days. If the nest lies under a sunny part of the beach and rain is minimal, egg temperatures can rise to 95°F (35°C), hatching the eggs in as few as 45 days. Therefore, eggs laid on the same night but in different parts of a beach might hatch weeks apart. Also affecting nest temperature is the heat generated by the metabolic processes of each clutch of eggs. During the latter half of incubation, temperatures in the clutch can be five to seven degrees warmer than the surrounding sand.

Temperature controls much more than the day of hatching. It also determines each hatchling's sex. Most vertebrates inherit sex chromosomes from their parents. The chromosome you inherit determines whether you are a male or female. In hu-

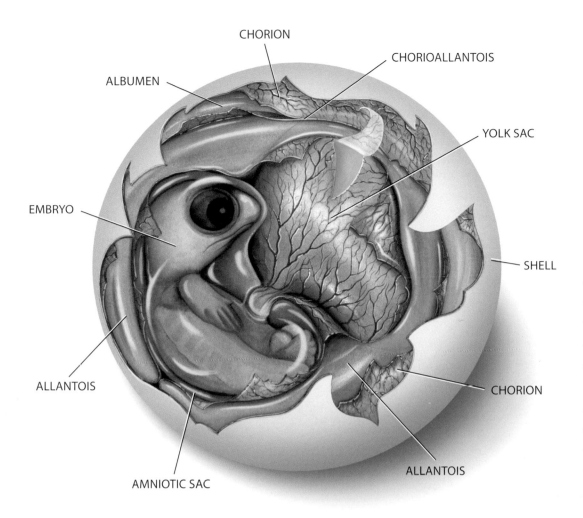

CHORION

CHORIOALLANTOIS

ALBUMEN

YOLK SAC

EMBRYO

SHELL

ALLANTOIS

CHORION

AMNIOTIC SAC

ALLANTOIS

A sea turtle embryo develops inside of an eggshell that keeps it from drying out but allows for gas exchange. As development proceeds the chorion and allantois fuse to form the chorioallantois membrane, which facilitates embryonic "breathing."

mans, two X chromosomes result in a female whereas one X and one Y chromosome cause the embryo to develop into a male. But sea turtles don't have sex chromosomes. Their sex is determined by the temperature at which the eggs incubate. For example, at 82°F (28°C) green turtle hatchlings develop into males, while at 88°F (31°C) the hatchlings become females. Temperatures between these two extremes produce clutches with a mix of both males and females. To complicate matters further, the heat produced by the eggs themselves affects the sex of the hatchlings. Eggs in the center of the nest are warmer and thus more likely to become females, while those on the edge of the clutch are cooler and more likely to become males. The first and last eggs to be laid, then, are more likely to become male sea turtles. Indeed, which way an egg bounces into the nest might determine its sex.

Many studies have shown that the middle third of incubation is a critical time for sex determination. Two days of rain during this period can lower the temperature of the nest four to six degrees. Thus eggs that were on their way to becoming females can end up as males merely because of passing storms.

Day by day the embryos are changing within the nest. Heads, eyes, limbs, and shells slowly take shape, as do internal organs. Each day, however, the risk of death is never far away. Besides human poaching, sea turtle eggs face a host of threats, including other mammals, insects, and crabs. Fungi can invade the nest, killing the eggs. If the nest becomes inundated with tidewaters for too long, the embryos will be robbed of oxygen and die. Yet those that survive these threats get no respite, for they must soon face a whole new set of challenges.

Entering the World

About two months after the eggs are laid they begin to hatch. First one and then another, the little turtles rip through their leathery shells using a pointed "egg tooth" on the top of their beaks. The movement of one hatchling seems to excite those in the surrounding eggs, setting them to squirming inside their shells. Soon they too have broken holes in their shells and set off their neighbors, and so forth. After the shell is pierced, the hatchling sits in its egg and absorbs what is left of the yolk sac until the turtle's plastron (bottom shell) straightens out. While waiting, the turtles seem to rest and build strength for the next stage of their odyssey. Turtles that hatch early may rest for as long as a day or two while their siblings hatch and absorb their yolks.

Once a large group of hatchlings have escaped their eggs, the group begins what is called "hatchling frenzy," crawling about and climbing over and bumping into one another. Hatchlings at the top of the heap knock down sand from the roof of the nest. This sand trickles down through the group and those on the bottom

Characteristics of Reproductive Female Sea Turtles

Average size, estimated age at first reproduction and time in years between nesting migrations of nesting female sea turtles.

Species	Carapace Length Inches (cm)	Age at First Reproduction (Years)	Remigration Interval (Years)
Green Turtle			
Southeastern U.S.	39 in (99 cm)	27-30	2.3
Australia	35 in (90 cm)	30-40	5.0
Hawaii	36 in (92 cm)	30-35	4.0
Caribbean	39 in (99 cm)	26-31	3.0
Hawksbill	31 in (79 cm)	20-25	2.9
Olive Ridley	26 in (66 cm)	11-16	1.7
Kemp's Ridley	25 in (65 cm)	11-16	1.5
Loggerhead			
Southeastern U.S.	36 in (92 cm)	28-33	3.0
Australia	38 in (96 cm)	28-33	3.8
Mediterranean	30 in (76 cm)	?	2.0
Oman	37 in (94 cm)	?	?
South Africa	34 in (87 cm)	17-30	2.6
Japan	33 in (84 cm)	?	?
Brazil	41 in (103 cm)	?	?
Leatherback			
Atlantic/Caribbean	60 in (153 cm)	9-15	2.3
Pacific	57 in (145 cm)	9-15	3.8
South Africa	63 in (160 cm)	9-15	3.0
Flatback	36 in (91 cm)	?	2.6

Sea Turtle Reproductive Effort during a Season

Average number of eggs in a clutch, number of clutches laid in a nesting season, time between clutches (internesting interval) and hatching success of nesting female sea turtles.

Species	Clutch Size	Number of Clutches	Internesting Interval	Hatching Success
Green Turtle	110	3	12-14 days	90%
Hawksbill	130	3	13-16 days	79-92%
Olive Ridley				
Arribada	110	2.2	17-45 days	2-10%*
Solitary	110	2	14 days	80%
Kemp's Ridley	110	1.5	20-28 days	81%
Loggerhead				
Southeastern U.S.	112	3.5	12-16 days	80%
Australia	127	3.4	14 days	87%
Mediterranean	97	?	13-17 days	?
South Africa	?	3.9	?	?
Brazil	127	?	?	73%
Leatherback				
Atlantic/Caribbean	85	7	9-10 days	53%
Pacific	65	7	9-10 days	54%
Flatback	54	2.8	16 days	79-95%

* Of all nests laid during an arribada

stamp it into the floor of the nest. As this continues, the nest rises slowly, like an elevator, toward the surface of the beach. The group effort makes climbing out much easier.

It takes 24 to 48 hours for the hatchlings to reach the surface, testifying to the labor involved. In general, the hatchlings travel up in groups of 20 to 120 individuals. But stragglers, generally late hatchers, can emerge up to two days after the first siblings leave the nest. Small groups often make it to the surface, but single hatchlings have great difficulty doing so. Some do manage to reach the air alone, but many become trapped part way up. They simply lack the strength to climb out.

There are three signs that a nest is about to disgorge its occupants. The first is smell. A few minutes before the hatchlings break through to the surface—even up to half an hour before—you can smell the scent of egg contents that clings to the body of the hatchlings as well as the smell of churned-up wet sand with its soil bacteria and fungi. Humans can detect the smell a foot or two away from the nest. Dogs and raccoons apparently can detect nests from greater distances. A study conducted in Tortuguero, Costa Rica, found that 33 percent of the local leatherback nests were attacked by dogs—all just as they were hatching.

The second sign is in the sand itself. A few minutes before the turtles reach the surface, the sand starts to sink above the nest. It drops only about an inch or two, but it is noticeable. When you see that depression form in the sand you know that the hatchlings are near.

A newly hatched olive ridley turtle pokes its nose out of the sand on a beach in Costa Rica. The first of a group of hatchlings to emerge from a nest, it will soon be joined by as many as one hundred brothers and/or sisters, the sex of the hatchlings being dependent upon the temperature of the nest during the middle third of development.

Technology is required to detect the third sign. We know that the temperature of an egg clutch peaks just as the hatchling frenzy starts. Scientists that study nesting turtles often insert temperature probes into the nests. As the hatchlings crawl up above the probe the nest temperature drops and you know they are on their way up. Any of these signs means that soon you will see a bunch of little heads popping up through the sand.

The Long Crawl

Just after sunset or during the night the hatchlings pull themselves out of the sand and onto the beach. Very, very rarely do they emerge during daylight hours, contrary to the impression given by popular nature shows and films depicting such behavior. Those scenes were probably staged. If hatchlings approach the surface during the day they get hot and stop moving. Of course if it is cloudy and rainy the sand can be cool enough for hatchlings to emerge in the daytime. As a rule, however, it is when the sun goes down and the sand cools off that the hatchlings finish their crawl to the surface. If they reach the surface during the night they just keep moving and boil right up out of the ground. Then it's off to the races. Getting out of the nest is usually a cooperative exercise, but crawling from the nest to the sea is a race, and it's every turtle for himself or herself.

The trip to the water can be as short as one yard for hatchlings emerging from a nest laid just above the high tide line and longer than a football field for those leaving at low tide. In both cases the hatchlings must dash to the sea before they are discovered by predators, a wide variety of which stalk the beach at night. In the Americas these include raccoons, coyotes, dogs, coatis, feral pigs, ghost crabs, and night herons.

When hatchlings emerge they enter a strange new world. Initially they are confused by cool air, wind, new smells, and the moon and stars. Before streaking for the surf they will sit still for a few seconds to several minutes, literally getting their bearings.

The cues that hatchlings use to find the sea are primarily visual. In the 1960s scientists David Ehrenfeld and Nicholas Mrosovsky conducted a series of exquisite experiments using simple tools but clever ideas to determine how hatchlings navigate on the beach. They covered the hatchlings' eyes by placing small black hoods over their heads and put them in little arenas on the beach. These turtles crawled in circles or headed off in random directions. Hatchlings with uncovered eyes were able to find the sea even it was over a small hill in the sand. Later experiments revealed that hatchlings find the sea because the sky is slightly brighter over water than over land.

If the sky is very dark and overcast or if it is raining very hard, hatchlings will use other cues to find the water. They tend to crawl down a slope, which will generally lead them toward the sea, and they are also drawn toward the sounds of the surf. The moon has little effect on hatchling orientation because hatchlings are most sensitive to light that is near the horizon.

Hatchlings, then, have both primary and backup systems for orienting their crawl to the sea. However, these instincts can spell danger for hatchlings

emerging from beaches near human development. Lights and noise from human dwellings can present false cues and actually lead the hatchlings away from the water.

Even a hatchling that knows where to go may well find the long crawl a difficult one. For the tiny hatchling, which is less than an inch tall, a branch washed up on shore becomes a fence; a driftwood log, a wall. They must climb over, under, and around many obstacles, and must do so before the sun comes up and birds can see the easy prey. With skill and a bit of luck the hatchlings will feel the roll of a wave that picks them up and carries them seaward.

Opposite: If this green turtle makes it through the surf off Surinam it still has less than a 50 percent chance of surviving through its first year. Years later it may return to the same beach, using a variety of navigation senses including an internal magnetic compass.

Left: Leatherback hatchling setting out to sea in the Huon Gulf of Papua New Guinea. It has "extra large" flippers and is counter shaded with white on its plastron so that its visibility to fish is reduced against the bright sky. The egg tooth that ripped open the eggshell is obvious as a projection on its nose.

Entering the Sea

Hatchlings that reach the surf line keep crawling because the first wave or two usually drags them backward and leaves them on the sand. Soon a bigger wave lifts them up and their flippers no longer touch sand. They begin to swim and at the same time dive toward the bottom, where they are captured by the undertow and swept out about five to ten yards from the beach into deeper water. They are now "sea turtles" in the true sense of the term. Females will not return to the beach for a decade or more, and the males will never be back.

Hold a newly emerged hatchling in your fingers and you can watch the swimming motion of its flippers—down and then up, similar to a bird in flight. The hatchlings paddle furiously as they try to escape from the beach. By now they are quite thirsty, having lost up to 20 percent of their body mass to evaporative water loss during their arduous climb out of the nest, the sprint across the beach, and frantic swimming against crashing waves and strong currents. They start drinking sea water and set a course for the open ocean by swimming in the direction of the oncoming waves. Over the next day or two the hatchlings will filter the salt out of the seawater they drink and replace the water that they lost. Hatchlings have large salt glands behind their eyes that excrete excess salt as tears. Without the glands, they would be dead in a matter of days.

Since waves normally approach the beach from directly offshore, swimming into the waves takes the hatchlings toward the open ocean. Sea turtles' brains contain magnetite (an iron compound). This means that hatchlings can sense the earth's magnetic field because the magnetite molecule is drawn to the North Pole like a needle in a compass and the brain cells around the magnetite can sense that pull. As they swim into the waves, then, hatchlings are able to detect the direction in which they are heading. When the hatchlings get far from shore, the waves no longer provide directional cues. Nevertheless, hatchlings continue on the same compass course they used when they left the beach. They probably use magnetic navigation from that point onward, whether moving about at sea or, many years later, finding their way back to the nesting beach where they hatched.

The Lost Years

The late Archie Carr, often called the "father" of sea turtle biology, wrote extensively about sea turtle mysteries. He was quite curious about the time from the moment a sea turtle entered the sea to when it was seen feeding along the coastline as a juvenile. Unable to measure that period of turtle development, Carr dubbed it "the lost year." When he died in 1987, the mystery lived on. Since then it has become clear that it takes not one but several years for the hatchlings to appear in their juvenile feeding grounds. Thus biologists began to call that time "the lost *years.*" Still, many questions remained.

Carr spawned a generation of scientists determined to solve the lost years puzzle. Now we believe we have a sense of what happens to sea turtles after they leave the beach. The answers didn't come cheaply, and many a researcher used novel methods that often required grueling fieldwork to find those answers. Research suggests that hatchlings spend the first few days of their lives swimming out to sea and feeding opportunistically on animals such as ctenophores (small jellyfish-like creatures) and tiny larval crabs and shrimp. They do not have to find a lot of food because they still carry some yolk inside their bodies. In fact, yolk will be a reserve supply of food for up to two weeks.

After a few days they are far from shore and beyond the reach of their worst predators. Fish of all kinds, pelicans, frigate birds, gulls, and terns patrol the surf and near-shore areas off the nesting beach. Any hatchlings still near the beach after daybreak usually end their brief lives as nutritious snacks.

Safer offshore, the hatchlings swim for only a few hours at a time. When not swimming, they fold their flippers along their sides and rest, floating in the ocean currents. Though they continue to head out into the open ocean, the strong currents eventually carry them to areas where ocean currents converge. There they find their

foraging grounds—drifting islands of seaweed rich with plants and small animals, including snails, small crabs, barnacles, fish larvae, ctenophores, small jellyfish, and shrimp. They also find ants, beetles, leafhoppers, and flies that blow out from land and find refuge in the seaweed islands.

We have a pretty good idea of where loggerhead hatchlings go after they leave their nesting beaches along the east coasts of Florida, Georgia, and South Carolina. In the late summer and early autumn the tiny turtles swim out to the Gulf Stream and drift along its edge, eventually meeting up with rafts of floating *Sargassum* seaweed—their new mobile homes. The *Sargassum* rafts drift eastward in the North Atlantic currents, often taking the little travelers off the coast of Portugal. From there they swing south past the Azores and toward the Canary Islands and Cape Verde Islands off the west coast of Africa. Then they turn westward, arriving back along the eastern seaboard of the United States. This circular migration of ocean currents, collectively called the North Atlantic gyre, can take years to complete. Eventually the hatchlings find themselves where Archie Carr predicted they might be: the Sargasso Sea, near Bermuda. There the hatchlings find food and a measure of safety in the midst of this floating seaweed archipelago in the eye of an oceanic whirlpool.

Juvenile loggerhead turtle raised in Florida by the National Marine Fisheries Service. Loggerheads in the Atlantic Ocean usually spend their early years riding the currents in the area of the Azores and Cape Verde Islands, hiding and feeding along patches of seaweed.

A threesome of mating green turtles is fully occupied on the Great Barrier Reef, Australia. The bottom turtle is the female and the middle turtle is a successfully mating male. The top turtle is a secondary male that seems to be either attempting to mate or pulling the other male off the female.

Other sea turtles undertake similar migrations but with some interesting variations. Hawksbill turtles also swim out to drifting seaweed, and then move to coral reef habitats within one to three years. Green turtles ride the North Atlantic gyre but appear to spend less time in it than loggerheads, and avoid floating seaweed altogether. Their distinctly bicolored bodies—white below and dark gray to black on top—suggest that they are open water animals. They eat anything that they can catch, including small oceanic snails, ctenophores, and other animals.

Leatherback hatchlings swim fairly continuously for at least six days, resting at night, and then disappear. We have no idea where they go once they get beyond coastal waters. They are not seen again until they are caught by fishermen in the open ocean perhaps three years later, their shells having grown in length to about 30 inches (76 cm). Flatback turtles also remain a mystery. Our best guess is that these Australian turtles do not enter the open ocean, instead remaining close to the northern coast of the island continent in shallow, somewhat muddy water.

Adolescence

After the initial lost years, most species of juvenile sea turtles move out of the open ocean and reenter coastal waters. By this time, the shells of those living along the east coast of the United States and in the Caribbean are about the size of dinner plates, while those in places like Hawaii and Australia are somewhat larger. These turtles are older and grow more slowly owing to a less reliable food supply. Feeding habits now become more like those of adults. Loggerheads and Kemp's ridleys begin to eat shellfish and crabs, and green turtles turn to algae or *Thalassia*, the latter commonly known as "turtle grass."

Hawksbills move onto shallow reefs, rock outcroppings, and into mangrove estuaries. In the Caribbean, juvenile hawksbills eat sponges primarily but also devour other animals such as sea cucumbers, sea anemones, and mollusks. Flatback turtles are found in the same shallow, turbid waters where we suspect the hatchlings all live. Where juvenile olive ridleys spend their lost years has long been a mystery, but recent observations suggest they live in the open sea in places where vegetation gathers.

Coming of Age

Sea turtles reach sexual maturity ranging from 5 years of age for captive green turtles on a high-quality diet at the Cayman Island Turtle Farm to perhaps 35 years or older for wild turtles in Australian and Hawaiian waters. Leatherbacks are the fastest growing wild sea turtles and mature at between 5 and 15 years of age, most at around 10.

As they age, juvenile males produce more and more testosterone. Their tails grow longer than the females', their plastrons soften, and a large hard nail on each front

Sea Turtles Nest at Different Times Around the World

Major nesting seasons for sea turtles in different regions of the world. In all of these areas some sea turtles may nest outside the main nesting season.

Species	Region	Season	Species	Region	Season
Green Turtle	Florida, Caribbean	May–August	**Kemp's Ridley**	Mexico	April–July
	Eastern Pacific	October–January			
	Australia	October–April	**Loggerhead**	Southeastern U.S.	May–August
	Arabian Gulf	April–September		Australia	October–March
	Malaysia	January–August		Eastern Mediterranean	June–August
				Brazil	September–February
Hawksbill	Caribbean	June–October			
	Yucatán	April–August	**Leatherback**	Caribbean	March–June
	Bahia, Brazil	October–March		Pacific	October–February
	Arabian Gulf	April–July			
	Seychelles	October–January	**Flatback**	Australia, Queensland	October–January
	Malaysia	January–August		Australia, Crab Island	June–October
	Indonesia, Australia	December–May		Australia, Northern Terr.	May–November
Olive Ridley	India	January–April			
	Mexico and Costa Rica	July–December			

flipper grows longer and starts to curve. On the outside, females simply get larger as they approach maturity, but internally their blood will contain increasing levels of both estrogen and testosterone.

As the first breeding season approaches, female production of estrogen decreases and testosterone production greatly increases. This apparently signals that it is time for the reproductive migration back to the nesting beach. A surge in their testosterone signals the males to start their migration toward the mating grounds, which usually lie just off the nesting beach or along the migration route.

Returning to the Beach

Extensive tagging efforts by Archie Carr and others in the mid-twentieth century demonstrated that females return to the same beach to nest year after year. Biologists began to suspect that sea turtles, like salmon, may even return to their own natal sites to nest. In the 1970s Australian scientists began to test this possibility by tagging large numbers of hatchlings—136,000 loggerheads and 109,000 green turtle hatchlings! However, only one immature green turtle and one immature loggerhead turtle have been recaptured from this effort. In the 1990s biologist Brian Bowen found a simpler method. He compared DNA from turtles on nesting beaches and discovered patterns of relatedness that make sense only if almost all females are returning to the beaches from which they hatched.

We suspect that adult sea turtles find their way back home using many methods, such as magnetic navigation, celestial cues, chemical concentrations in sea water, and memory of landmarks. However they do so, most sea turtles find their way across hundreds and even thousands of miles of ocean to the sands from which they crawled as hatchlings up to three decades before. Simply finding the same beach twice is remarkable, let alone finding the beach from which you hatched and returning year after year for decades.

Finding a Mate

Male sea turtles begin their reproductive migration before females. They race to the mating grounds and eagerly wait for the females to appear. At the sight of a female the males burst into action, swimming after her and immediately attempting to crawl up onto her carapace (shell) to begin mating. If the female accepts a male he mounts her quickly. Using the claws on his front flippers to hold onto the leading edge of the female's carapace, he twists his long tail under that of the female to insert sperm into her cloaca. They can stay in the mating embrace for hours both near the surface and at the bottom. Occasionally they surface to breathe but then go right back down.

Mating is not a gentle affair for the female. Males often bite females on the flippers, neck, and head, leaving open sores that can take weeks to heal. It is no picnic for a male, either. Other males often hang around the mating pair, and they will do what they can—ram him and bite him—to try to break up his embrace with the female. If he is dislodged another male quickly replaces him. Females do control partner selection, however. If she does not approve of the switch she covers her cloaca with her hind flippers and swims rapidly to the bottom where she presses herself tightly to

Temperature Dependent Sex Determination (TSD) in Sea Turtles

Low temperatures during incubation of eggs produce males and high temperatures produce females. The temperature that produces 50:50 ratio of male to female hatchlings is called the pivotal temperature.

Species	Pivotal Temperature	100% Male	100% Female
Green Turtle			
Costa Rica	85°F (29.5°C)	<82.5 (28.0°C)	>87°F (30.5°C)
Suriname	84°F (28.8°C)	82°F (27.8°C)	86°F (30.0°C)
Hawksbill			
Brazil	85°F (29.6°C)	83°F (28.4°C)	87°F (30.4°C)
Antigua	84.5°F (29.2°C)	83°F (28.5°C)	85.5°F (29.8°C)
Olive Ridley			
Costa Rica	87°F (30.5°C)	80.5°F (27.0°C)	89.5°F (32.0°C)
Kemp's Ridley			
Mexico	86.5°F (30.2°C)	<84°F (29.0°C)	88°F (31.0°C)
Loggerhead			
Southeastern U.S.	85°F (29.5°C)	<81.5°F (27.5°C)	87°F (30.4°C)
Australia	83°F (28.2°C)	79°F (26.0°C)	88°F (31.0°C)
Australia	84°F (29.0°C)	<79°F (26.0°C)	88°F (31.0°C)
Brazil	84.5°F (29.2°C)	82.5°F (28.0°C)	87°F (30.6°C)
South Africa	85.5°F (29.7°C)		
Mediterranean	84°F (29.0°C)		
Leatherback			
Costa Rica (Pacific)	85°F (29.4°C)	84°F (29.0°C)	86°F (30.0°C)
Suriname	85°F (29.5°C)	83.5°F (28.7°C)	85.5°F (29.8°C)
Flatback	85°F (29.5°C)	84°F (29.0°C)	89.5°F (32.0°C)

the sand. Thus protected, males can only try to coax her into approving them or wait until she needs air and harass her on her trip to the surface. Females are receptive to mating only during the month before the nesting season and the 12 hours after they lay each clutch of eggs.

Back to the Beach

When a female goes ashore to lay eggs for the first time in her life, she is entering a world wholly different from the one she has known for the previous 15 to 35 years. We do not know if she senses the magnitude of the moment or remembers being there as a newly emerged hatchling. In prehistoric times that beach likely would be just as it had been 35 years earlier. Today that beach might have changed considerably, its palms and beach grasses displaced by hotels or houses with patios and concrete pools. But the adult female seems largely undeterred by the changes, the urge to lay her eggs overwhelming. If she spots people before she nests, she will likely turn around, go back into the sea, and return later.

If the beach is quiet and the female is undisturbed she crawls up the beach for a few minutes. Then she pauses and takes a couple of breaths. After looking around she continues inland until she finds the right spot to dig her nest. First she prepares the ground by clearing away any vegetation or debris with broad sweeps of her front flippers. Then she digs a body pit, using her front flippers to sweep out an area about the size of her body. Depth ranges from a few inches for a ridley to a foot or more for green turtles.

With the pit completed, she digs her nest completely with her hind flippers. Unable to see her work, it appears that she uses only her sense of touch. At first she digs a cup-sized pile of sand with one hind flipper and throws it out to the side. Then she sets that flipper on the side and reaches in with her other flipper to repeat the process—right, left, right, left. This behavior is so programmed that if one flipper is missing, the stump of the leg goes through the same motion, the other flipper waiting for it to finish each cycle before continuing.

Scooping sand out by the handful, the female creates a deep, flask-shaped hole. She scrapes the edges of the nest chamber and carves it into shape. Finally a flipper comes up with no sand. After a couple of empty swipes she stops and grows quiet. Then—plunk, plunk, plunk—those golf-ball-sized eggs begin to drop.

Once all of the eggs are laid, the female starts to cover the nest, again with her hind flippers. In an alternating process she scoops in sand from the right and left to fill the top of the nest and the neck of the nest hole. She pats the sand down after each scoop. When the hole is filled she continues to tamp down the sand and to test its compactness with her tail until the nest is covered and only the pit remains. After a brief rest she throws sand over the area with her front flippers for as long as it takes to totally obscure the nest. She continues to throw sand as she slowly moves away from the nest, leveling out the surrounding area. This done, she heads for the surf, completing the cycle and launching another generation of sea turtles on its way. ⌒

A green turtle covers her eggs on Sangalakki Island in Borneo as the sun begins to light up the morning sky. This final stage of the nesting process can take 20 minutes or more. She probably started constructing her nest in the dark hours of early morning, resulting in her finishing as sunrise approaches.

CHAPTER THREE

BIOLOGY
Under the Shell

THE BIOLOGY OF SEA TURTLES IS FASCINATING. THEIR SIZE, SHAPE, TEMperature, salt control, and other factors combine to make them unique animals. When we see the image of a sea turtle gliding through clear water our first thought is, "That is a beautiful animal." Naturally, the mechanics—the 'how'—behind the beauty of the image are far from our mind, yet the anatomy and physiology of sea turtles are what underlie the image we see. Exploring the parts of the sea turtle explain volumes about the animal as a whole.

Anatomy

We start our journey with the shell, where we encounter large deposits of bone. In all species except the leatherback the shell is made of scutes, which are pieces of keratin covering the bone. Keratin is a hard protein substance that is remarkably similar to the protein that forms our fingernails and hair.

The carapace, or dorsal (top) shell, forms while the turtle is still in the egg. As the embryo develops, the ribs expand and then fuse to the vertebral column. The ribs then grow connections to one another, forming a hard, shield-like structure. The plastron, or ventral (bottom) shell, also develops within the egg, when a series of "plastron bones" spreads out under the skin in an area that equates to the human chest. Plastron bones appear to be unique to turtles, not some modification of a common structure as we see with the ribs becoming the carapace. The same group of embryonic cells that form the skull during early development are also the source of plastron bones.

Unlike other sea turtles' hard shells, leatherback shells have no keratinized scutes but rather a leathery skin that lies over a lose mixture of thin bony plates connected by soft cartilage. Under the leathery shell lies a thick layer of oily fat and fibrous tissue. Several bony keels, or ridges, run along the back of the turtle from the neck area to the tail. These keels appear to help in swimming, perhaps one of the factors contributing to the fact that the leatherback is the fastest swimming sea turtle. The leatherback's skin is quite smooth, more like that of a dolphin than of a cow. As hatchlings, leatherbacks' shells are covered with small scales, providing some added protection.

Previous pages: The plastron, or lower shell, of a sea turtle is formed from bones that arise during embryonic development (from the same set of embryonic cells that form the skull). As with hatchlings, this hawksbill's white coloration seems to camouflage the turtle from sharks lurking below.

Opposite: A green turtle swims toward the surface off Fiji with both a remora on its plastron and a diver on its carapace.

Like the human skull, sea turtle skulls are composed of thick bones. A principal difference is that there are no soft openings in the turtle's temple regions. The sea turtle brain is relatively small compared to body size, but some interesting adaptations allow the animal to survive for long periods of time in oxygen-poor environments. Unlike some species of terrestrial and freshwater turtles, sea turtles cannot retract their heads into their shells.

Instead of teeth sea turtles grow horny beaks made of keratin. The shape of the beak varies by species and appears to reflect diet. In hawksbills it is pointed like the beak of a hawk, adapted to bite off pieces of sponge. In green turtles it is serrated like a knife for cutting sea grass. In loggerheads and ridleys the beak is larger and thicker, allowing these turtles to eat hard-shelled animals like clams and crabs. Leatherback beaks have sharp points on both the upper and lower jaws so that they can slice into jellyfish. Flatbacks have a more generalized beak and eat a variety of soft and hard-shelled animals.

In most regards the internal organs of turtles resemble those of other vertebrates, including a chambered heart (three as opposed to the four chambers found in humans), two lungs, a lobed liver, stomach and digestive system, arteries and veins, and the usual bones and muscles in the legs. The appendages, however, differ significantly from freshwater turtles. Sea turtle limbs are highly specialized as flippers, allowing them to swim much faster than their freshwater cousins. They also allow the turtles to take long voyages in search of food and their nesting beaches. However, the flippers are vulnerable to attack by predators, and it is not unusual to find a turtle with a flipper wound or a missing limb.

The upper bones in the legs (humerus and femur) are shortened and thickened, when compared to those of a lizard or crocodile, and held close to the body. The lower bones (radius/ulna and tibia/fibula) are also shortened and proportionately smaller than those of other reptiles. Likewise, the wrists and ankles are flattened and expanded whereas the fingers and toes are elongated and form most of the flattened and expanded part of the flippers. The front flippers form a wide, long, wing-like structure and the hind flippers resemble a large hand or baseball catcher's mitt. The front flippers help the turtle fly through the water and the hind flippers act as a rudder for steering.

The Hot Blood of Turtles and Dinosaurs

The biology of sea turtles has a lot to tell us about how dinosaurs and early marine reptiles lived. Sea turtles share with dinosaurs their large size and long-distance migrations. The giant marine reptiles, such as pleisiosaurs, swam in the oceans during the Age of the Dinosaurs more than 100 million years ago, facing many of the same challenges as early sea turtles. Even the land dinosaurs share common ground with modern sea turtles. All of these animals were large, lacked the mammals' fur coat, and presumably did not have the heat-generating metabolic processes of both mammals and birds.

Dinosaurs have always fascinated people. Big, bold, and with long necks or huge heads and teeth, they thrill us without scaring us, perhaps because they are from an age long gone. In recent years we have developed new views of their exciting lives, new drawings of muscular giants migrating across the plains of ancient earth, and new ways of looking at their biology. Dinosaurs have also become "stars." My own hometown of Haddonfield, New Jersey, erected a bronze statue of *Hadrosaurus*, the duck-billed dinosaur which was discovered there.

Recently it was suggested that dinosaurs were "hot-blooded" like mammals. The reasoning was that they must have had to move quickly, like mammals, so they must have had high metabolic rates. Being warm-blooded means that an animal generates lots of heat internally by its own metabolism. Cold-blooded animals, on the other hand, obtain heat from the environment, generally from the sun, water, warm rocks, and so forth. However, animals such as lizards can be quite warm after basking in the sun a few minutes, despite the fact that they are considered cold-blooded.

When the hot-blooded dinosaur stories hit the newspapers I was studying animal heat transfer and temperature regulation with the world's expert in the field of biophysical ecology, Dr. David Gates. David read the articles and said to me, "This doesn't make sense. Why don't you do some calculations and see if it could really

The sharp, serrated beak of a green turtle from the South Caicos Islands, British West Indies. The beak works like a set of garden shears by cutting off turtle grass and other algae upon which the green turtle feeds. The beak is made of keratin, a protein similar to that found in human fingernails.

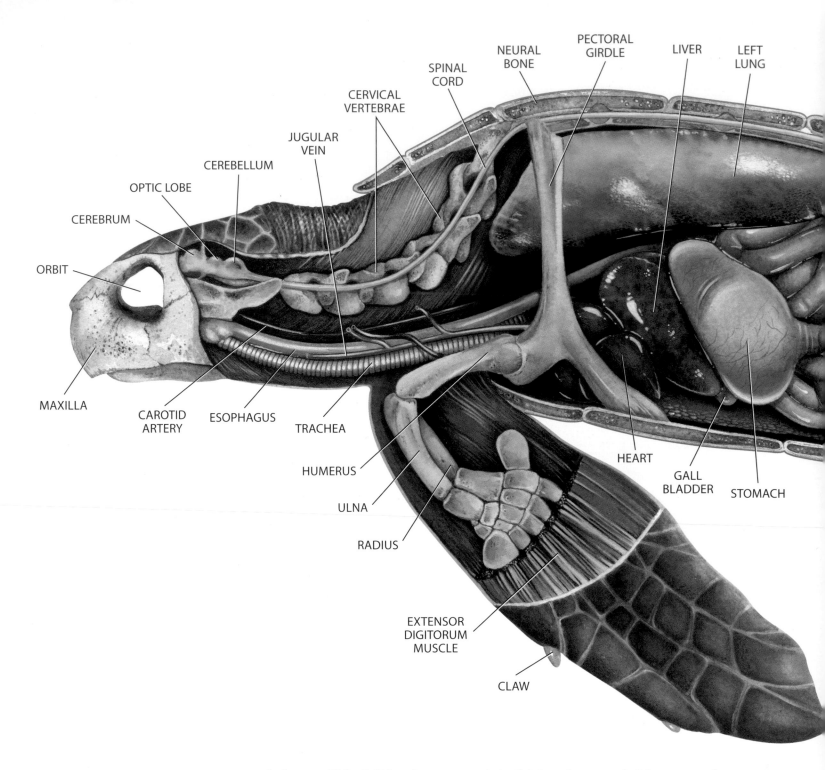

ORBIT

CEREBRUM

OPTIC LOBE

CEREBELLUM

JUGULAR VEIN

CERVICAL VERTEBRAE

SPINAL CORD

NEURAL BONE

PECTORAL GIRDLE

LIVER

LEFT LUNG

MAXILLA

CAROTID ARTERY

ESOPHAGUS

TRACHEA

HUMERUS

ULNA

RADIUS

HEART

GALL BLADDER

STOMACH

EXTENSOR DIGITORUM MUSCLE

CLAW

work that way?" So I did and sure enough it *didn't* make sense. I did some mathematical modeling with some colleagues and determined that a dinosaur in the warm Mesozoic Era would have had a nice warm body temperature of 95°F (35°C) even if it had no metabolism at all. It might, I determined, need to move into the shade on occasion to cool down, but it certainly didn't need metabolic heat to have a high body temperature. In truth, dinosaurs had to be cold-blooded or they would overheat given their large body size.

Some from the hot-blooded camp complained about our results, saying that all we had were mathematical models. We needed to study the body of a really big rep-

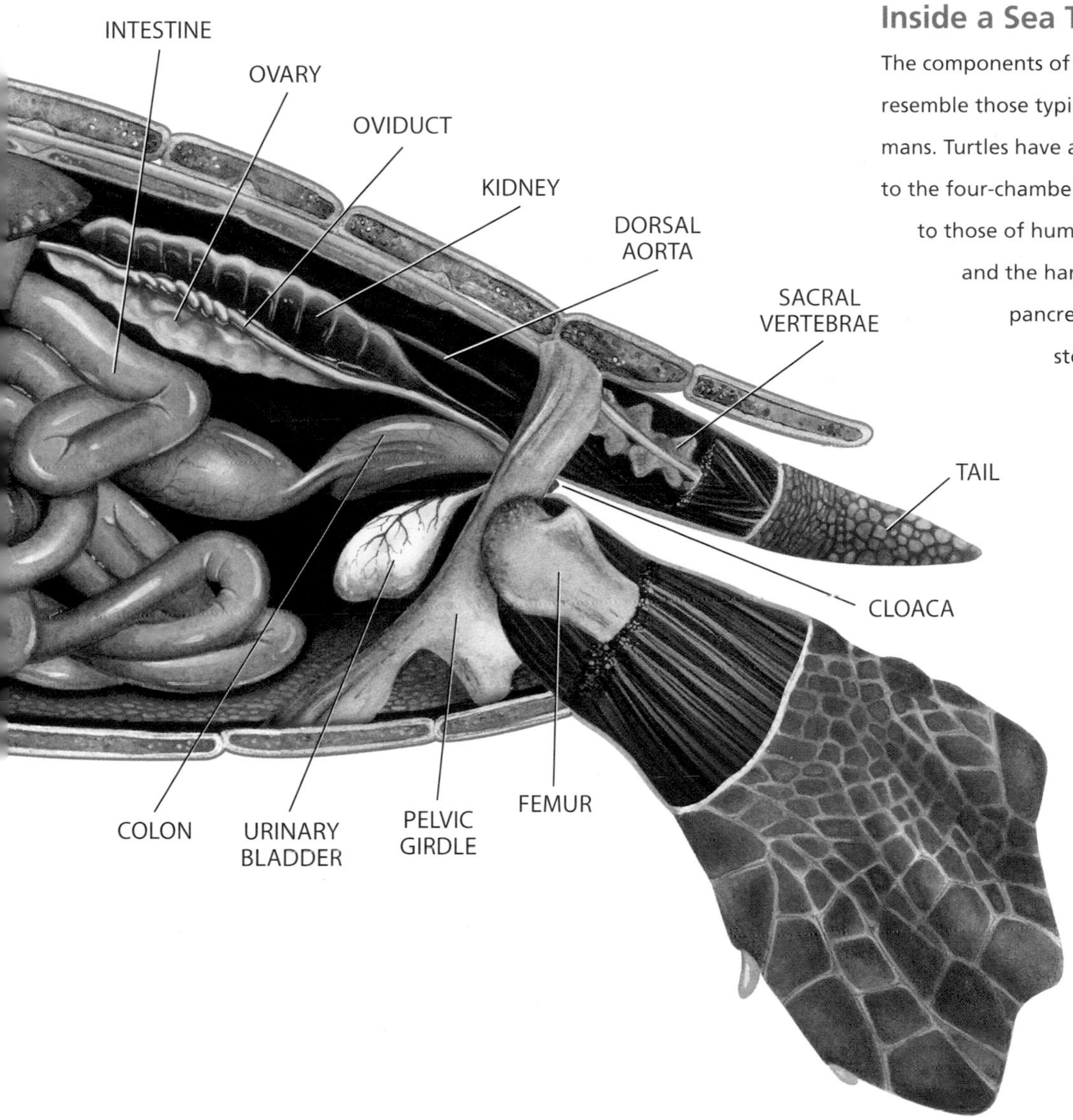

INTESTINE
OVARY
OVIDUCT
KIDNEY
DORSAL
AORTA
SACRAL
VERTEBRAE
TAIL
CLOACA
FEMUR
COLON
URINARY
BLADDER
PELVIC
GIRDLE

Inside a Sea Turtle

The components of the anatomy of a sea turtle generally resemble those typically seen in vertebrates including humans. Turtles have a three-chambered heart as compared to the four-chambered heart seen in mammals. Compared to those of humans, the limb bones are shortened and the hand and foot bones are elongated. The pancreas and spleen are mingled with the stomach and intestines and thus cannot be seen in the drawing.

tile to see if it had a high metabolic rate. If modern day reptiles could be highly active with a low metabolic rate then our statements about dinosaurs would be supported.

What reptile could we use as a model of a dinosaur? Large crocodiles don't move around very much and they eat people, making them somewhat difficult study subjects. Sea turtles, however, can be extremely large, swim great distances, are very active—and don't eat people. We began our research with green turtles. Measuring the body temperatures of adult females on land and in the ocean, we quickly found that they often had a high body temperature but did not have the high metabolic rate that would have labeled them warm-blooded. Still our critics were not satisfied. They suggested a larger reptile was needed for comparisons, so we focused our studies on leatherback turtle metabolism.

Leatherbacks are twice the size of green turtles, swim in very cold water, have low metabolic rates typical of cold-blooded animals, and yet their body temperatures are relatively high. Indeed the term "cold-blooded" seemed oddly out of fashion. There appeared to be many ways to have warm blood, and not only like mammals and birds.

Technology

To measure the body temperature of a swimming green turtle we had to apply new technology and do something that many said was impossible—successfully track a turtle in the ocean with radio transmitters. Biologist Archie Carr had written about all the efforts over many years to track sea turtles with various technologies, and even of getting the National Aeronautics and Space Administration involved in the process. It didn't work. Nowhere was it more evident than in Tortuguero, Costa Rica, where the magnetic, black sand beach made contemporary electronics go haywire. Not knowing any better, I set out in 1978 with two of my students, Ed Standora and Bob Foley, to do what others said was impossible.

Ed was an electronic genius who was fascinated by animals. He had spent his youth building ham radios and learning how to put transistors together. For our studies he designed a nifty multichannel sonic transmitter that could measure the body temperature of a turtle in several internal locations and send the data as sound waves through the water to our recording devices. He also developed new radio transmitters, which he said would work only when the turtle came to the surface because radio waves are absorbed by seawater. Trying to minimize the discomfort, we attached Ed's devices to green turtles in the shade on the beach at Tortuguero. We placed the radio transmitter on a small hydrodynamic float that the turtle towed behind her, so it would be up out of the water when she broke the surface. Then we released her and she made her way into the waves, disappearing into the sea.

Lacking funds for a proper boat, we commandeered an old rubber raft from a storage shed. Apparently it was left over from a French expedition. We pumped it up with air and attached an old outboard motor. I was to stay on shore while Ed and Bob stayed close enough to the turtle to receive the sonic and radio signals. They put all of their gear in a waterproof box and launched the raft through the surf. The motor quickly died and soon a wave capsized the raft in the surf.

On the beach I could see disaster brewing: I was fairly sure I had seen a bull shark in their vicinity. Immediately I called to them to return to the beach. Ed looked back at me with uncertainty, and I realized he was trying to find Bob. Thirty seconds passed—still no sign of Bob. Finally, his head broke the surface. He had gotten tangled up in the raft. They were able to swim to shore, none worse for the wear. Their waterproof box and the raft washed up a little ways down the beach.

I bravely told Ed and Bob to gather their things and try again. They looked at me with shock, but soon realized it was the only option. Ed muttered something about it being "easy for you to say" but began cleaning the salt water out of the engine as I helped strain the sand out of the gasoline. We launched the boat again and this time I cheered as they made it beyond the breakers. Soon they heard a loud hiss and realized the raft had a tear. They quickly slapped on some duct tape, pumped some more air into the raft, and set about finding the turtle. She had not traveled too far, so they located her within a few minutes. She could have easily slipped beyond our reach.

Soon they disappeared from sight, so I just sat and waited. Hours passed; day turned to night. Finally I heard the faint sound of a motor. I put out a searchlight and sent up a flare. The raft looked like it had lost half of its air. They rode the waves onto the beach, their gasoline nearly gone, exclaiming triumphantly that they had done it.

With the help of Archie Carr, the telemetry raft is launched at Tortuguero, Costa Rica. Researcher Bob Foley holds his hand on the throttle, watching for a break in the waves so that he can "gun the motor" and make it to smoother waters.

We had been the first to successfully track sea turtles using radio transmitters. I am often given credit for it, but my students deserve the glory. They followed the first two rules of our expeditions: Bring everyone back alive and bring back the data!

That data indicated that the green turtle did indeed have warm blood when she was swimming vigorously. Her thick shell and large mass retained the heat produced by her exercise, working just like our mathematical model. She was warm, but not warm-blooded.

Cold Stunning

Every year in New England and during cold winters in Florida, something happens to sea turtles that is unusual in the animal world and alarming to humans. The cold water actually causes sea turtles to float to the surface, unable to swim. It's called "cold stunning," and it occurs when the ocean temperature falls to about 45-50°F (7-10°C). Cold-stunned turtles are seen near the end of October through early December along Cape Cod, Massachusetts, and Long Island, New York. Recent studies suggest that these turtles are generally young and arrive in the late spring and early summer to feast on the abundant crabs. A small percentage of these turtles stay too long and fail to migrate south before the onset of colder temperatures. In Florida, particularly in Indian Lagoon near Cape Canaveral, cold-stunning occurs occasionally in January when a cold front passes through the central part of the state. Sea turtles are also found cold-stunned in the Gulf of Mexico and along the shores of Western Europe. Green turtles, loggerheads, and Kemp's ridleys seem particularly susceptible. Smaller turtles are affected sooner than larger ones, the latter taking longer to cool.

A fibropapilloma tumor covers the eye of a green turtle in Hawaii. There is no cure for the infection, but in some cases surgery can ease the burden and stop the growth of tumors, perhaps by activating the immune system.

Having lost their ability to swim and dive in the cold, the stunned turtles float like beach balls bobbing on the waves. Often they are blown by the wind and wash up on beaches. Those of us brave enough to walk the beaches during extremely cold weather occasionally encounter a live but very limp sea turtle. Turtles exposed to the cold for too long will die, but if they are rescued in time and warmed they often recover. Some turtles in this condition need the services of a wildlife veterinarian. Most government resource agencies in affected areas are now set up to assist sea turtles that have succumbed to cold stunning.

Disease

We tend to think of sea turtles as swimming free in pristine oceans and therefore generally free of disease. There are, however, no more pristine oceans, and a spectrum of environmental stresses including pollution, temperature, salinity changes, and physical trauma all reduce a turtle's immune response to disease. Just as with humans, when a turtle is stressed its adrenal glands produce corticosteroid hormones that lower the immune system's ability to fight disease.

Wild sea turtles seldom suffer from malnutrition. If an emaciated turtle is found it is usually because it has been injured or has some other medical problem preventing it from feeding. It is rare for wild sea turtles to suffer from bone diseases, iron deficiencies, or bacterial, fungal, and viral infections. Yet captive animals are susceptible to all of these diseases. Wild sea turtles carry a variable load of parasites such as leeches, barnacles, trematodes (primitive flatworms that live in the digestive system, heart, and blood vessels), and nematodes (roundworms that live in the digestive system). However, these problems pale in comparison to the greatest threat to sea turtles worldwide—cutaneous fibropapillomatosis, or fibropapilloma for short. This disease causes large wart-like masses to grow on the turtle's skin.

Green turtles have been hardest hit by fibropapilloma, but hawksbills, loggerheads, flatbacks, olive ridleys, and leatherbacks also can be affected. First discovered in 1938 in captive green turtles from Florida, the disease was then found among Hawaiian green turtles in 1958. In 1978 we saw a green turtle at Tortuguero with small fibropapilloma tumors, but since the disease was not widely known at that time we thought it merely an oddity. In the 1980s, fibropapilloma increased rapidly and became an epidemic in many locations, especially among Hawaiian green turtles.

Now fibropapilloma occurs in every ocean basin. Most at risk are turtles living in near-shore waters, in areas next to large human populations with poor sewage treatment facilities, and in lagoons and other areas with low water turnover. In some areas, up to 92 percent of the turtle population is infected. Apparently water pollution adds excessive nutrients and disease agents into the sea. These stressors depress the turtles' immune systems and then they succumb to the disease.

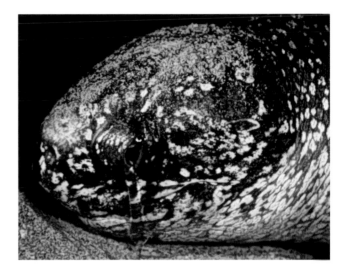

Afflicted turtles develop growths on their skin and in internal organs. Skin growths may be smooth or ulcerated and generally resemble cauliflower. They can be small, or large enough to obstruct the turtle's eyes and mouth. Especially large, warty masses on a flipper can make swimming difficult. These growths can also occur on the neck, base of the tail, and under the legs.

Internal growths occur as firm white nodules in the lungs, stomach, intestines, and kidneys. Eventually these are, of course, lethal. The external growths can be removed by careful surgery and are not malignant, so they do not grow back if removed. Freezing off small-to-medium growths with liquid nitrogen appears to stimulate the immune system to fight off other untreated growths as well.

There is no cure for fibropapilloma and the cause is not known definitively. One hypothesis is that a herpes virus is responsible, but the available data are insufficient for drawing firm conclusions. Of great concern is the rapid spread of fibropapilloma from green turtles to other species, and from Hawaii and Florida to many other locations around the world.

Salt of the Ocean

Shipwrecked sailors who drink salt water quickly die because the ocean contains 3.5 percent salt while our blood contains only 0.9 percent. The salt water literally sucks water out of our systems and the cells in our body shrivel up and die. Sea turtles live in and drink the ocean. Salt water does not harm them because they have special adaptations to reduce the intake of salt water with their food and special glands to excrete excess salt from their blood. The sea turtle's esophagus, which connects the mouth to its stomach, is specially adapted to eliminate salt water taken in with food.

All sea turtles have a large salt gland behind each eye. This modified tear gland has two functions. First it concentrates and excretes salt, preventing the turtle from experiencing the ravaging effects of excess salt in the body. Second, the tears produced by these glands lubricate the eye when the turtle is on land laying its eggs and throwing sand over the nest.

When a sea turtle eats food, whether it is a leatherback eating jellyfish or a green turtle eating sea grass, it fills its esophagus and then contracts the muscles around it so that excess water that came in with the food is expelled out of the mouth. Lining the esophagus are hundreds of long, closely packed, cone-shaped structures called papillae, which consist of connective tissue covered with keratin. Papillae are about an inch long and have a thick, hard, outer layer and point backward toward the stomach. When the esophagus contracts, the food is held in place by the papillae and water is pushed out. The sea turtle literally has to "burp" water after eating.

Another adaptation developed because the turtles' kidneys are unable to concentrate the sea turtle's urine sufficiently to get rid of the excess salt. To prevent overdosing, sea turtles have evolved a salt gland. This large, modified tear gland is near the eye. Other sea animals have similar adaptations. Sea snakes, for example, have modified salivary glands that excrete excess salt. In marine lizards, nasal glands do that job. Saltwater crocodiles have evolved glands on their tongues to excrete salt.

A sea turtle's salt gland occupies a large area in its head. The openings of the tear ducts continuously excrete a thick, clear, highly salty mucous that lubricates the eyes and eliminates excess salt at the same time. The impression people get when they see this is that the sea turtle "cries," the sight of which has inspired numerous myths. Some suggest that sea turtles "weep" as they nest because they know they will not see their babies again or because their hatchlings will have to face the myriad hazards of life at sea.

The salt gland produces a large volume of fluid that is twice as concentrated with salt as seawater. Thus a sea turtle that drinks one quart of seawater will eliminate the excess salt by secreting half a quart, or a pint, of tears that is about 6 percent salt. The remaining pint of freshwater is used for a variety of functions. Hatchling salt glands are proportionately larger than those in adults. When they emerge from the nest they are dehydrated, having lost about 12 percent of their body mass through evaporation during their climb up out of the nest. The enhanced salt gland allows them to drink large volumes of seawater and quickly replace that water loss.

Diving

Sea turtles are deep divers, and for this they have evolved a physiology that is more similar to diving mammals than other reptiles. Indeed, a typical sea turtle spends 95 percent of its time underwater and about one hour a day, cumulatively, at the surface. In warm conditions, sea turtles normally spend 15 to 20 minutes at a time underwater searching for food. However, they can remain active underwater for 45 minutes or more without breathing. At rest, they spend more time submerged and sometimes sleep straight through the night submerged. In winter they can even hibernate in the mud, as happens in Florida's Cape Canaveral ship channel and in Mexico's Gulf of California.

Diving Abilities of Sea Turtles

Sea turtles are excellent divers. They can dive to great depths and remain submerged for extended periods of time. Their diving prowess rivals that of the deepest diving mammals.

Species	Depth: Ft	m	Pressure: lbs/in²	Duration (min.): normal	maximum
Sea Turtles					
Green Turtle	360	110	162[a]	9-23	66
Loggerhead	757	233	342	15-30	60
Olive Ridley	940	290	426	28-40	45
Kemp's Ridley	150	50	74	12-18	45
Leatherback	3900	1200	1764	5-20	45
Hawksbill	325	100	147	56	74
Flatback[b]	488	150	220	?	?
Mammals					
Human	198	61	90	1	2.5
Walrus	260	80	118	10	–
Sea Lion	812	250	368	25	–
Gray Seal	325	100	147	20	–
Weddell Seal	1950	600	882	43	–
Sperm Whale	3686	1134	1667	75	–
Blue Whale	325	100	147	43	–
Harbor Porpoise	65	20	29	12	–
Bottle-nosed Porpoise	975	300	441	28	–

[a] Pressure increases at a rate of 1 atm/10 m of water depth. 1 atm equals 14.7 lbs/in²

[b] Estimated based on depth of water where flatbacks live and feed on the bottom

Before a deep dive the turtle approaches the surface and begins to exhale, blowing bubbles into the water just before its head pops out. Then it takes a big breath, or several, and goes back under, descending until the carbon dioxide buildup must be purged and the turtle must resurface for a refill of oxygen. Leatherbacks hold the record for the deepest dive among sea turtles. Using depth-sensing transmitters, we recorded a dive of 2,417 feet (737 m) by a female leatherback migrating from Costa Rica to the Galápagos Islands. Another female was recorded at an estimated depth of 3,248 feet (990 m) near St. Croix in the U.S. Virgin Islands. Less spectacular but still quite deep, an olive ridley was recorded at 940 feet (287 m) and a loggerhead at 754 feet (230 m). In the Pacific Ocean off California, a green turtle was photographed at a depth of 360 feet (110 m), its plastron pushed inward by the pressure at that depth. The flexible plastron, then, absorbs water pressure, which explains why the deep-diving leatherback has a softer, more flexible shell. Indeed, an underwater video revealed a leatherback's plastron bending inward at a depth of just 90 feet (27 m).

Sea turtles can quickly empty and refill their lungs because their air passages are reinforced with cartilage and surrounded by smooth muscle. When a turtle reaches depths of 260-325 feet (79-99 m) its lungs collapse, leaving air in the reinforced air passages. Because these passages do not have the extensive blood vessel arrangement that is present in the lungs, turtles are protected from the dangerous condition known as the bends. Human divers get the bends when nitrogen is forced from the lungs into the blood during long and/or deep dives. When the diver surfaces too

quickly, nitrogen forms bubbles in the blood, disrupting the transfer of oxygen and carbon dioxide within the body and causing excruciating pain. Sea turtles' unique lungs and the restriction of blood flow to them during a deep dive spare them the agony of bends.

Basking

On a sunny day in spring it is common to see large numbers of freshwater turtles basking on logs and other objects, soaking up the sun. This behavior warms the turtle to more than water temperature, allowing for increased activity levels. Basking is common among alligators and crocodiles but was long thought not to occur in sea turtles. However, recent observations suggest that sea turtles do bask in places not usually visited by people or predators. Green turtles bask in the Galápagos Islands, on French Frigate Shoals in the Hawaiian Islands, and in Australia. The common features of these sites are isolation, cool-to-cold ocean temperatures (72°F/22°C or lower), and steady winds. Turtles warm up to a shell temperature of about 104°F (40°C) and then flip sand onto their backs and flippers. This helps to cool their skin by as much as 20°F/11°C, allowing the body to continue heating up without burning the skin. Many of the basking turtles are females, possibly warming up so their eggs mature more quickly.

Sea turtles also bask in the water. Olive ridleys float at the surface in the Pacific. They appear to be asleep but are actually warming themselves. Studies suggest a loggerhead can warm up a few degrees higher than water temperature when basking at the surface with most of its carapace exposed to the sun. Their lungs are attached to the muscles just below the carapace, so it is possible that the large number of blood vessels in the lungs help deliver warm blood to the rest of the body.

Flying through the Oceans

Sea turtles do not swim like freshwater turtles. The local pond turtle swims by alternately paddling its limbs in a diagonal sequence—first right front and left rear legs, then left front and right rear. A sea turtle swims by simultaneously sweeping its front limbs through the water. This movement is quite similar to the flapping of a bird's wings, and creates a figure-eight pattern when viewed from the side. The front flippers move up and out and then down and in together in a power stroke that provides propulsion forward and up or down depending upon how the flippers are oriented. There is little thrust when the turtle starts to move its flippers forward, but it is carried forward by momentum. When the flippers approach their maximum extension they rotate and provide thrust by producing lift similar to that caused by air flowing over the top of a wing. Then, as the turtle twists its flippers and starts to sweep them back, additional forward thrust is generated again through lift. Since lift works perpendicular to the orientation of the flipper, the turtle moves forward.

Previous pages: A green turtle swimming in sprinkles of sunlight. Green turtles can dive to depths of over 360 feet (110 m) and can stay submerged for an hour or more. Although normal dives usually last less than 20 minutes, green turtles sometimes sleep under water all night with their heads tucked under coral reefs.

As the flipper goes through the second half of its retraction, it provides forward thrust by pushing against the water like a paddle. In this phase the flippers encounter resistance, or drag, as they push back against the water. The turtle glides forward by momentum as it starts to move its flippers forward and the process repeats itself. Sea turtles turn by changing the amount of sweep of one of the front flippers and by rudder-like actions by the hind flippers. Most sea turtle hatchlings swim by dog-paddling, whereby all four limbs act as paddles, and by rear flipper kicking, in which the rear flippers beat backward and the front flippers are held close to the body. Hatchling leatherbacks, on the other hand, have long foreflippers and swim very much like adults.

Green turtles literally fly through the water using their front flippers to provide both lift and thrust just like the wing of a bird.

Walking

Sea turtles use several different styles of walking. The hatchlings of most species walk by alternately using diagonally opposite limbs. They push back with one front flipper and the opposite rear flipper while bringing the other two flippers forward. Then they retract those flippers and bring the first two forward. Adult loggerheads, hawksbills, olive ridleys, and Kemp's ridleys alternately move their front flippers and move their hind flippers together or alternately. Both hatchling and adult leatherbacks crawl by moving the front and rear flippers at the same time. Green turtle hatchlings use the diagonal motion, but adults walk like a leatherback.

Magnetic Personalities

Sea turtles accomplish amazing feats of navigation and appear to do so, in part, by using magnetic information. Many green turtles that feed along the coasts of Africa and Brazil converge on Ascension Island in the middle of the South Atlantic to lay their eggs. Other green turtles feeding off Brazil lay their eggs on beaches to the north, in Surinam. Leatherbacks feeding off Nova Scotia migrate to St. Croix in the U.S. Virgin Islands to lay their eggs, while leatherbacks feeding off Chile in the South Pacific migrate to Costa Rica or Mexico to lay their eggs. Apparently sea turtles can determine their position relative to a goal, that is, they have a map sense. It seems that adult turtles remember where they have been and know where to go next to carry out their life functions.

Hatchling turtles orient themselves toward the ocean primarily by visual cues on the beach. In the surf, they swim against the direction of the waves. Once they get out of the surf zone, they orient themselves by using an internal magnetic compass. Although more studies are needed to confirm how adults migrate, they also may use their magnetic compass to navigate in the open seas.

Sea turtles probably detect magnetic information by using the magnetic forces acting on the magnetite crystals in their brains. In this way they can detect the inclination of the earth's magnetic field, that is, the angle at which the magnetic field intersects the surface of the earth. This angle is essentially zero degrees of inclination at the equator and 90 degrees at the poles. They can also detect the intensity or strength of the magnetic field, which varies across the surface of the earth. With this information they may be able to virtually map the oceans magnetically. This would allow them to navigate precisely toward a particular goal, such as their nesting beach. But if their magnetic sense is disrupted they can use other senses such as sight and smell to assist them in their migrations.

Sight and Smell

Sea turtles have good vision underwater and can see colors. However, they are most sensitive to light from violet to yellow (wavelengths of 400-600 nanometers). They do not see light in the orange-to-red portion of the visible spectrum (630-700 nm). This is not surprising because seawater absorbs the longer wavelengths of light. On land sea turtles are nearsighted because the lenses of their eyes are adjusted for the refraction caused by water and because their spherical shape and rigidity do not allow for adjustments out of the water.

Three bones in the human middle ear, the malleus, incus, and stapes, allow us to hear a wide variety of sounds both in air and under the water. Humans hear sounds in air in the frequency range of 20-20,000 Hertz (Hz). Sea turtles have

only one ear bone, called the columella, and do not have an external ear. Their tympanum, or eardrum, is a continuation of the skin on the skull. Sound is conducted through the tympanum and a layer of fat to the columella and then to the inner ear. The amount of sound pressure needed to move the columella increases with the frequency of the sound, so a sea turtle hears only low-frequency sounds. Green turtles detect sounds from 200-700 Hz, loggerheads from 250-1,000 Hz. This works well for sea turtles underwater but they cannot hear very well on land. However, sound can also be conducted to the inner ear directly through the bones of the head, backbone, and shell, allowing sea turtles to sense vibrations when on land. This might explain how a sea turtle can sense that people are walking up behind it on a beach when it cannot see them.

A sea turtle's nose, like our nose, has external openings and connects to the roof of the mouth through internal openings. It contains two sets of structures that allow it to smell, one each on the upper and lower surfaces of the nasal passage. The olfactory nerve connects to both areas. The lower surface has two sets of sensory cells collectively called the Jacobson's organ. Sea turtles share this structure with snakes. After sampling the air with its tongue, a snake then puts its tongue into its Jacobson's organ to sense chemicals in the air—that is, to smell. A sea turtle can smell under-

Having cut its way out of the leathery shell, this hatchling green turtle next expects to tunnel up toward the surface. Hatchlings start to flap their flippers as soon as they break out of the egg. They keep flapping and swimming for up to six or seven days with only short breaks at night when they fold their flippers against their body to rest.

most turtles, crocodiles and alligators, some lizards, and the tuatara, a lizard-like reptile of New Zealand. As a rule in turtles, low temperatures produce males and high temperatures produce females. We presume that temperature affects one of the turtle's genes by turning it either on or off. Some of the genes involved in human sex determination have been found in turtles. However, since there are no set sex chromosomes it is not clear how these genes are turned on and off. What is the switch? There is still much to discover about how this process works.

This seems like a very haphazard way for an animal to establish its sex, which might result in lopsided sex ratios in a turtle population. For example, one might imagine that a particularly cool decade would produce very few females, or that in an especially warm period there might not be enough males to fertilize all the eggs or maintain genetic diversity in the population. Indeed, in a given year many beaches do produce a female-biased sex ratio among the hatchlings. At Playa Grande in Costa Rica we found that from 1993 to 1996 leatherback nests ranged from 74.3-100 percent female. In fact, when we checked the weather records we concluded that in most years since 1950 the percentage was likely to be the same. Only two or three years were cool enough to produce a large number of male hatchlings and a sex ratio closer to 50:50. These male hatchlings were produced primarily during the rainy season when the sand was cooler.

We were concerned that there might be a shortage of males in the leatherback population and that many eggs were not being fertilized. That could explain the low hatching success in this and perhaps other leatherback populations. However, when we checked the paternity of the hatchlings we found that in 30 percent of the clutches more than one father had mated with the mother, suggesting that males were not in short supply. In another study we discovered that the fertility rate was 90-100 percent. Finally, in yet another study we attached an underwater video camera to the carapaces of females to observe mating behavior. There were generally two and three males chasing each female soon after they were released into the water. Despite the female-biased sex ratio among the hatchlings, it appears there are plenty of adult males around to fertilize the eggs.

Loggerheads in South Carolina and Georgia produce about 56 percent females from their nests, but in Florida they produce 87-100 percent females. Likewise in the Mediterranean and Brazil, the loggerhead hatchling sex ratios are female-biased. Green turtles in the Mediterranean, Costa Rica, Surinam, and Malaysia also produce more female than male hatchlings. Hawksbills in the Caribbean and Brazil produce mostly female hatchlings, as do olive ridley turtles on the Pacific coast of Costa Rica. However, some green turtle nesting beaches in Australia produce more males than females and different beaches on Ascension Island produce very different sex ratios owing to differences in sand temperature. The question of why there appears to be, in general, a sex-ratio bias remains a mystery. Ideas such as a warming of the average earth temperature abound, but we are far from definitive answers. ⌒

Kemp's ridley hatchlings race to the sea at Rancho Nuevo, Mexico.

CHAPTER FOUR

HISTORY
The Ancient Lineage

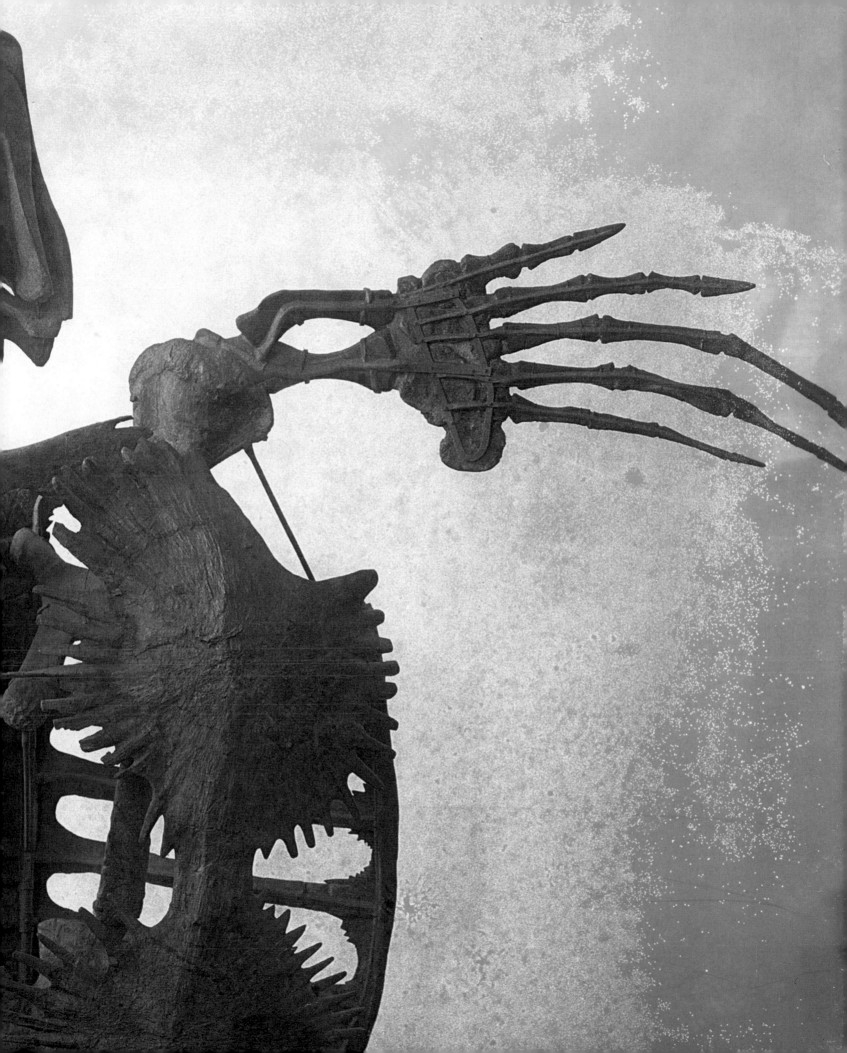

THE FIRST REPTILES AROSE IN THE PENNSYLVANIAN PERIOD ABOUT 300 million years ago. These animals looked like big stocky lizards and gave rise to many groups, including the mammals, crocodiles, dinosaurs, marine reptiles, lizards, and turtles. The first turtle fossils lie in 220-million-year-old sediments from the Triassic Period in Germany and in sediments from about the same time in Greenland and Thailand. The distribution of these fossils, even considering the subsequent movement of continents, suggests that turtles were widespread during this era. They became successful in the Triassic and in most regards have not changed much since then.

If you had walked along the edge of a lake 200 million years ago and encountered one of the early turtles, you would have seen *Proganochelys*, an animal that looked much like a snapping turtle does today. *Proganochelys* is the earliest known fossil turtle. It was large for a freshwater turtle, approaching three feet in total length with a shell about two feet long. It had the classic components of a modern turtle—an obvious carapace and plastron, and a skull with a horny beak. A few small teeth remained in the roof of the mouth, but they were soon lost as subsequent generations proceeded through the evolutionary process.

The scutes on the carapace of *Proganochelys* were raised into points and the neck and tail had small spikes, all of which probably offered some much-needed protection. *Proganochelys* lived in shallow, swampy lakes with phytosaurs (ancient crocodile-like reptiles), labyrinthodonts (six-foot-long amphibians that looked like oversized stocky salamanders), and predatory fish. Prowling the edges of swamps were plateosaurs (early dinosaurs that walked upright on two legs).

Proganochelys literally pops into the fossil record as a completely formed turtle. Its immediate ancestors are unknown and there is no known transitional animal between other reptiles and turtles. Given that early turtles lived in the water and their shells would readily fossilize in the mud, it is puzzling that no fossils of turtle precursors exist. The hard shell of a turtle enveloped in muddy sediments without oxygen would have been preserved and eventually would have turned to stone. Theories and debates abound. Some scientists think turtles arose in the late Permian Period, about 260 million years ago, directly from the ancient stem reptiles that

Previous pages: *Archelon* was one of the largest sea turtles that ever lived. With a total length of up to 15 feet (4.6 m) and a head more than 3 feet (1 m) long, *Archelon* had a flipper span of 16 feet (4.9 m). This specimen is in the Peabody Museum at Yale University.

Opposite: Painting of *Proganochelys* during the Triassic, by Frank Ippolito. *Proganochelys* was an early turtle that lived in freshwater swamps at the same time that early dinosaurs and primitive fishes inhabited the earth.

Geological Timescale

Era	Period	Epoch	Age (millions of years ago)	Major Biological Events
Cenozoic	Quaternary	Recent	0.01	Historical time
		Pleistocene	1.8	Ice ages; humans appeared
	Tertiary	Pliocene	5	Apelike ancestors of humans evolved
		Miocene	23	Continued radiation of mammals
		Oligocene	35	Origin of many primate groups
		Eocene	57	Angiosperms and mammals dominated
		Paleocene	65	Major radiation of birds and mammals
Mesozoic	Cretaceous		145	Angiosperms first appeared; mass extinction of many organisms including dinosaurs; modern sea turtles appear
	Jurassic		206	Dinosaurs were abundant and diverse; first birds evolved; early sea turtles
	Triassic		245	Gymnosperms dominated landscape; radiation of dinosaurs; evolution of first true mammals; first turtles appear
Paleozoic	Permian		290	Radiation of reptiles; mammal-like reptiles evolved and most modern insect orders; mass extinction of marine organisms and amphibians
	Carboniferous		360	Extensive forests; first seed plants; amphibians dominant and first reptiles evolved
	Devonian		410	Diversification of bony fishes; first amphibians and insects evolved
	Silurian		440	Diversity of jawless fishes; first jawed fishes and diversification of early vascular plants
	Ordovician		510	Colonization of land by plants and arthropods
	Cambrian		540	Radiation of most modern animal phyla; origin of the earliest vertebrates occurred
Precambrian			4,600	Earliest traces of life; the oldest fossils of cells; the evolution of invertebrates occurred

looked like small lizards or dinosaurs. Others suggest a relationship with a later group of reptiles.

A Quick Change

The absence of fossil links is probably best explained by the theory that turtles arose very quickly in the Triassic Period and that their immediate ancestors did not live in water so they did not make good fossils. But what could have caused such a rapid change in the reptile line? The most recent research in molecular biology suggests that a change in only one or two genes that control the development of bone cells in the body could have led to the formation of the turtle shell. We know that there are a group of genes, called HOX genes, which control major pathways of development in an embryo. Loss of function by one such gene, for example, led to the loss of teeth in birds (and probably turtles as well). An increase in function of one or two controlling HOX genes may have caused the formation of the carapace and plastron that define the turtle. Since the shell of *Proganochelys* is very similar to that of a modern turtle, it is likely that the change from a stocky, lizard-like animal to a turtle happened in one or two evolutionary leaps.

Sea Turtle Relationships

Modern sea turtles arose in the Cretaceous Period and are the descendants of an ancient line of marine turtles. Two of the four families of modern sea turtles are alive today.

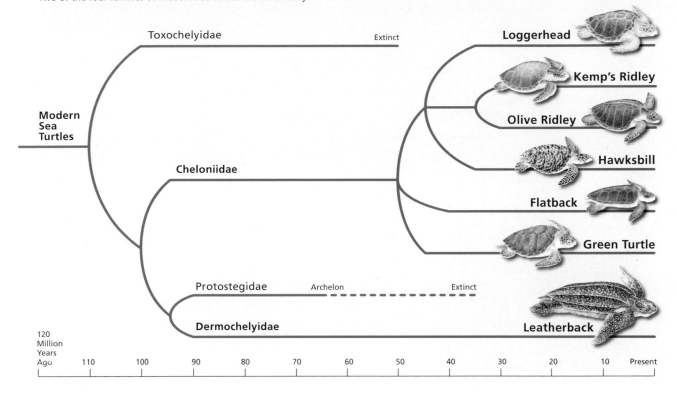

The added protection of the shell was a great advantage for turtles and most species have not experienced much evolutionary tinkering with that part of their body. The internal structures and the limbs, neck, and head have changed considerably, but the shell remains the same today as it was 200 million years ago.

The First Sea Turtles

Turtles moved into and out of the oceans several times beginning in the Jurassic Period and into the Cretaceous. These explorers had body forms more like modern freshwater turtles than today's sea turtles, but displayed varying degrees of change, an example being that their limbs formed into flippers. Through evolutionary time, these turtle species came and went. Some were widespread and very successful, but eventually their lineages died out owing to changing environmental conditions or competition with other aquatic animals. By the end of the Cretaceous, the only turtles left in the ocean were from a new lineage.

Modern sea turtles arose about 110 million years ago in the ancient oceans. The oldest fossil of a sea turtle descendant is *Santanachelys gaffneyi*, a specimen found in eastern Brazil from the Early Cretaceous Period about 110 million years ago. It is a fossil of a small sea turtle, about the size of a dinner plate. It seems to have been a transitional body form, looking somewhat like a sea turtle and somewhat like a freshwater species. Well on the evolutionary path toward the sea turtles we see today, *S. gaffneyi* had relatively primitive, paddle-like front limbs and, unlike its descendants, it still had movable toes. Like today's sea turtles, it had a hard shell and large head as well as something no sea turtle can live without—large spaces in the skull that housed salt glands.

The ocean that this little turtle entered was a forbidding place, filled with predatory fishes, sharks, and now-extinct reptilian predators such as ichthyosaurs, mosasaurs, and plesiosaurs. It was a tough, dangerous world, but *S. gaffneyi* had the advantage of a shell as hard as a rock. *S. gaffneyi* made it through that treacherous beginning and eventually diversified into four distinct types of sea turtles, each sufficiently different that taxonomists have classified them as the four families of prehistoric sea turtles.

All four families were swimming, feeding, and laying eggs on that fateful day some 65 million years ago when a giant asteroid struck the earth near Mexico's Yucatán Peninsula. There is little doubt that the result of that collision was a cascade of events that ultimately changed the earth's climate and led to the extinction of many species. Dinosaurs, which were already declining, soon disappeared, as did many marine invertebrates. While birds and mammals began their march toward domination over the land, a shocking 85 percent of the animal species inhabiting the earth vanished soon after the asteroid hit.

Sea turtles, however, hardly seemed to notice. In fact, they increased in diversity. They swam, ate, mated, laid their eggs, and tried to grow large enough that even sharks would find them a difficult meal.

Above: *Toxochelys* was an early sea turtle that measured about 24 inches (60 cm) long and had a broad shell. Openings in the bones of the carapace made the turtle lighter and more maneuverable. The shell was covered with muscle and skin. This fossil can be seen at the Smithsonian Institution in Washington, D.C.

Opposite: *Protostega* lived in the late Cretaceous approximately 80 million years ago. This fossil is in the Smithsonian Institution.

Surviving in the Ancient Seas

Two of the four families of sea turtles that occupied the oceans of the Cretaceous have since become extinct. The family Toxochelyidae, consisting of nine to ten types of small- to medium-sized sea turtles with broad, circular shells, diverged from the main line of hard-shelled sea turtles in the early Cretaceous. Over time they diversified into many different-looking animals, some of which specialized to live near the coast and others that adapted for the open oceans. Toxochelyidae descendants were abundant in the Great Inland Sea that covered the Great Plains of North America. They also flourished along the East Asian littoral and in the early Atlantic Ocean, which was forming as North America and Europe were spreading apart.

The Toxochelyidae diversifications took on many physical forms. Some species had a solid carapace and others a more reduced shell with large openings between the bones. These openings were covered with skin and perhaps horny scutes. With a smaller, lighter shell these turtles would have been faster swimmers than other species. Some turtles developed a secondary palate like that present in mammals. The secondary palate is the roof of the mouth that separates the air passages from the nose and mouth allowing an animal to breathe and chew food at the same time. It would have allowed these turtles to chew hunks of food while surfacing to breathe.

The Toxochelyidae turtles shared the oceans with modern hard-shelled sea turtles until the Eocene Period some 50 million years ago. Over time this group developed specialized feeding habits that may have been their ultimate undoing. They could not compete with their more generalized sea turtle relatives, which were able to change food sources as the climate and environment changed.

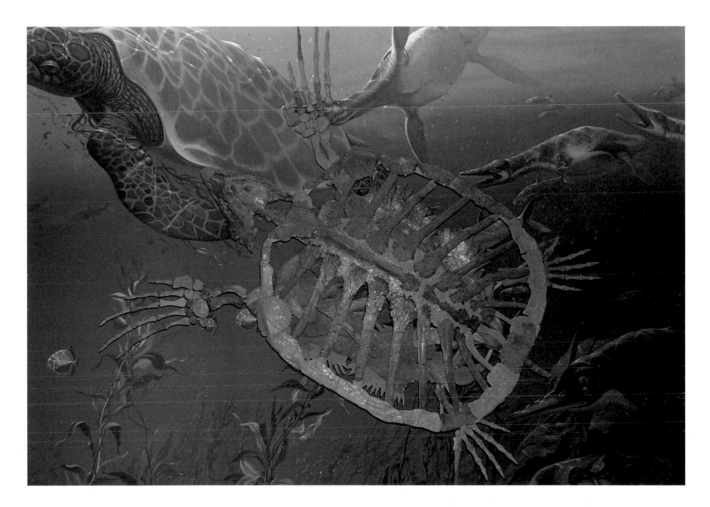

Another turtle family, Protostegidae, started with little *Santanachelys* in the early Cretaceous but gave rise to gigantic turtles later in the period. The most famous of these was *Archelon*, which grew to 15 feet (4.6 m), weighed over 6,000 pounds (2,700 kg), and its flippers spanned 16 feet (4.9 m). Its head was 3.3 feet (1 m) long and featured a sharply hooked beak. Living in the Inland Sea, *Archelon* probably ate large-shelled invertebrates such as clams, ammonites, and the ancient nautilus. Rooting around on the bottom, it pulled its prey from the sand and mud with its beak and crushed them with its massive jaws. Its carapace was only partially covered with bone

Archelon was enormous. The plastron was made up of large spiny, bones that, unlike those in modern turtles, did not fuse together. Giant flippers and a reduced carapace made this turtle very maneuverable. This particular animal lost a hind flipper, perhaps to a giant shark.

and had very large open spaces between the ribs. This reduced its weight, made it faster and more maneuverable, and helped it dive to great depths in a manner similar to the modern leatherback. The plastron was formed from several star-shaped pieces of bone. The Protostegidae all but died out with the dinosaurs; only one member of the family lived on into the Oligocene Period.

Rise of the Modern Families

Six of the seven species of living sea turtles belong to the third family, Cheloniidae. These turtles all have a well-developed, bony skull with a secondary palate and their carapace and plastron are fully formed of bone. They occupy a variety of ecological roles or niches in the oceans by living in a variety of habitats and eating different types of foods such as seagrass, crabs, clams, and sponges. Their ancestors arose in the early Cretaceous in parallel with the family Toxochelyidae. As many as twenty different types plied the oceans over time, but only four of these survive today as distinct genera, *Chelonia*, *Eretmochelys*, *Lepidochelys,* and *Caretta*. The Chelonids were more generalized species than the Toxochelids and thus were able to out-compete them for resources and were more adaptable to changing conditions in the oceans. During the Paleocene and Eocene periods the Toxochelyidae disappeared.

Finally there is the family Dermochelyidae, which has a sole extant species, the leatherback turtle. Splitting off from the Protostegidae in the late Cretaceous about 100 million years ago, the leatherback family reached a peak of diversity in the Eocene about 50 million years ago after the Protostegids all but disappeared. It was as if new niches opened in the seas and leatherbacks rushed to fill them. At least six different species of leatherbacks lived in the oceans worldwide, their fossils found in Africa, Europe, North America, and New Zealand. These turtles were very much like the modern leatherback, though carapace structure varied from fairly thin, flat shells to thicker, more robust shells with pointed ridges. All of them had shells made up of small bony plates.

As time progressed, competition likely increased with other animals like modern fishes and marine mammals. Indeed, the different leatherback species probably competed among themselves. Food sources likely became more limited and leatherbacks had to specialize as food types diminished. Leatherback diversity decreased to the extent that only two species survived at the end of the Miocene Period, and by two million years ago only *Dermochelys coriacea*, the modern leatherback, survived. It appears that the secret to our leatherback's success was its adaptation into a jellyfish predator. The leatherback had a large food supply all to itself, as the other species seemed to avoid the jellyfishes. With its specialized anatomy and great physiological adaptations it roamed the oceans making a unique living and increasing in numbers.

In the Days of Humans

In more recent times, millions of sea turtles filled the oceans. Reports by early European sailors indicate that, even as late as the eighteenth century, ships that had lost their way to the Cayman Islands in the Caribbean Sea could steer entirely by the noise generated by green turtles swimming there to lay their eggs. Columbus discovered the Cayman Islands in 1503 and named them Las Tortugas because the ocean was virtually wall-to-wall with green turtles. Ships were constantly bumping into them. Scientists conservatively estimate that there were 100 million green turtles in the Caribbean Sea alone before Europeans arrived there. They ate 13.4 trillion pounds (6.09 trillion kg) of sea grass a year, or about half the Caribbean's annual production of grass biomass. As the grass was consumed, nutrients were recycled and the grass beds were fertilized, resulting in a smoothly working ecosystem. The turtles seemed to completely control which plants and animals lived there.

Extrapolating these estimates from the Caribbean to other areas is risky, but considering that sea turtles were present in all of the tropical oceans, one gets a sense

that swims in a large ocean kicking up water with its flippers. Because the elephants walked in circles the world rotated during the day, and when the elephants tripped there were earthquakes. When the turtle splashed its flippers it created monsoons.

This Hindu myth even reaches into modern western society. In his book *A Brief History of Time*, author Stephen Hawking tells one version of a well-known story about an old woman and the famous English scientist and philosopher Bertrand Russell. Other versions employ other scientists and philosophers such as William James but tell the same story. The scientist has just given a magnificent lecture on the origin of the universe in which he convincingly described in great detail how the earth revolves around the sun and the sun in turn moves through a galaxy of stars. When he asked for questions an elderly lady in the back of the room got up and said, "That is all rubbish. Everyone knows that the world is a flat plate that rests on elephants on the back of a giant turtle." The scientist smiled and gave a knowing glance to his fellows in the first row before saying, "That is very interesting, but if your theory is correct, what is the turtle standing on?"

"On the back of a second and larger turtle," she replied.

"But what does that rest on?"

"Sir, you are a very clever young man, but not very wise," said the lady. "Everyone knows that it is turtles all the way down!"

Before Columbus

Although evidence of the impact Europeans had on sea turtles in the New World abounds, it is also clear that Native Americans had a significant effect on many sea turtle populations before the landings of Columbus. Very few such populations were untouched by humans before the first sailing ships arrived from Europe. Native Americans used tortoiseshell from hawksbill carapaces to make ceremonial and ornamental objects. The famous Hopewell Indian mounds, built 2,500 years ago as far inland as Ohio and Illinois, reveal tortoise shell pins, combs, hair pins, and other craft objects. Wooden statues of animals feature tortoiseshell eyes and bird wings. The shell and scutes of hawksbills were also important religious objects and were traded throughout the region.

Indeed, sea turtle remains are found in at least 23 sites in the United States ranging in age from 5,000 to 2,500 years. These sites contain remains of green turtles, hawksbills, Kemp's ridleys, and loggerheads. Native Americans not only ate sea turtles but also placed carapaces, bones, and ornamental objects made from tortoiseshell into burial sites. In the Caribbean region there are more than 40 sites containing sea turtle remains and at least 11 on the Yucatán Peninsula.

Caribbean peoples took turtles not only from nesting beaches but also from feeding grounds. Sea turtles were important components of the diet and culture in many of these societies. People on the smaller islands, Cuba, and on the Miskito coast of Nicaragua all traded in tortoiseshell. The relative contribution of sea turtles to the remains at some sites decreased through time, indicating that these people reduced the populations of turtles that they hunted. In parts of the Caribbean, aboriginals wiped out entire populations.

Mayan communities on the coastal areas of the Yucatán and Belize ate sea turtles and used their shells for containers and shields. Some communities like Isla Cerritos supplied turtle meat to larger population centers such as Chichén Itzá some sixty miles inland. Sea turtles played a symbolic role in the cultures of the Maya and other peoples of Central America, and are commonly represented in Mayan ceramics, figurines, and stone altars. Gold jewelry was fashioned into the images of turtles throughout the region. Ancient people used sea turtles in Argentina, Venezuela, Ecuador, Peru, and Chile, but the record of their importance in the local cultures is not as strong as that among the Maya.

Consider, for example, Costa Rica's granite and sedimentary rock balls. Ranging in size from a few inches to more than six feet in diameter, these balls are found in the Río Térraba valley on the west coast and in various pre-Columbian burial sites. They are almost perfect spheres and for years people have speculated about their significance. Some have suggested that they were religious symbols, others that they were items of commerce or symbols of wealth. The spheres were found on raised platforms and in rows, and sometimes were associated with burial grounds. There is, however, no consensus of opinion as to what they represent.

Yet to a sea turtle biologist the answer is simple. Ancient peoples were just as amazed by the prodigious reproductive effort of sea turtles as we are today. Sea turtles laid large numbers of eggs on the beaches of Costa Rica, providing a substantial and predictable food supply for local people. The simplest explanation for those large, perfectly round, white spheres is that they represent sea turtle eggs. All ancient cultures suffered from the need to produce enough offspring to overcome losses from disease, starvation, wild animals, and war. People needed symbols to boost their confidence in the bounty of nature and their own reproductive powers. Sea turtle eggs were undoubtedly an aphrodisiac then as they are today. Between 1,000 and 1,500 years ago the Chibcha people of Costa Rica were simply imitating the reproductive excess of the sea turtles that they saw on a regular basis. The giant stone sea turtle eggs were symbolic offerings people made to thank their gods, not only for the bounty brought ashore by the sea turtles but also for the increase in machismo gained by eating those eggs. The Chibcha had sea turtle eggs and we have fancy red sports cars and sport utility vehicles. Which culture is the wiser?

This granite sphere (moved to a location at the entrance of Liberia Airport in Guanacaste, Costa Rica) is typical of what some believe to be ancient turtle egg sculptures. The artisans who produced these symbols seemed to be paying homage to the reproductive output of sea turtles. Large and small, these "eggs" can be found at many locations throughout Costa Rica.

It is clear that since prehistory and antiquity humans have had a great effect on sea turtles. As biologist Jack Frazier points out, sea turtle populations were changed by humans before European technology arrived on the scene. Ancient peoples, seeking to feed their growing populations, affected sea turtle numbers, densities, and geographic distributions. They also affected the location and timing of behaviors such as feeding, mating, and reproduction. One example of a behavioral effect is the fact that sea turtles bask only on deserted islands uninhabited by humans. European expansion obviously escalated the pressures on sea turtles, even if their numbers were down from what they had been 20,000 years before the westerners arrived.

Food for Mariners

The abundance of green turtles was the main reason that Europeans were able to explore and exploit the Caribbean region. Exploration, settlement, pirating, and war all depended upon a large and dependable supply of food. Green turtles congregated offshore in the Cayman Islands to mate and came ashore in astonishing numbers to lay eggs from June to September. Sailors walked the beaches at night and turned the nesting females onto their backs. In the morning they went back and loaded them into small boats, then returned to their ships. The holds were filled with green turtles laid on their backs where they could survive for months without food. The fresh and tasty meat was also fortified with vitamin C, reducing the sailors' incidence of scurvy.

It seemed for a long time that the supply of turtles was limitless, but as we now know it wasn't. The first turtle nesting colony, or rookery, to disappear was on Bermuda. The next was on the shores of the Greater Antilles. The Bahamas soon followed and then the Florida Keys. The Dry Tortugas, originally named for the many green turtles that nested there, were emptied of their namesake. Finally, sometime in the 1800s, the Cayman Island population disappeared as well.

Efforts to find turtles for the big ships moved to the south shore of Cuba and, when those turtles were depleted, to the vast turtle grounds off Nicaragua's Miskito coast. Eventually the Nicaraguan government closed off the hunt to foreign vessels in the last half of the twentieth century. Fresh or salted, dried or pickled, green turtle meat fed an entire region and was a staple for the crews of ships from Great Britain and Spain. Turtles also helped open up the Caribbean to western influence, but the great turtle fleets passed silently away before the modern conservation movement raised an alarm. In those times the use of turtles was what we would call today "a matter of national security." Powerful nations needed sea turtles as they set out to conquer far away lands. So even had the conservation alarms been sounded, few would have heard and fewer still would have listened.

The hawksbill turtle was abundant in the Caribbean when Europeans arrived. Indeed, Columbus watched Native Americans hunting hawksbills along the southeast coast of Cuba in 1494. Hawksbill tortoiseshell was a significant trade item. As European trade and settlements expanded in the nineteenth century, large amounts were

exported to New York, Britain, and France. Japan used tortoiseshell for centuries in its *bekko* crafts, reducing populations of hawksbills in the western Pacific Ocean. It is likely that the advent of plastic substitutes for tortoiseshell saved the hawksbills from extinction.

As meat and products from sea turtles gained value beyond indigenous markets, sea turtles entered the cash society. Indigenous peoples spent more time hunting turtles for money and less time on traditional subsistence activities. With cash they bought nets and thus turtles became rarer still. The introduction of a cash value for turtles also added a new tension into the social structure of villages as some people became wealthier than others. Catching turtles soon became the path to wealth.

The effects of western expansion on sea turtle populations in the Caribbean were repeated in other parts of the world. As sailing ships reached new areas, sea turtles were a ready source of food. They appeared to be inexhaustible. For decades, even hundreds of years in some areas, sea turtles fed the world's navies and extended their sailing ranges. Without that food supply the spread of European civilization would have been much slower. When we think of the many good things that have resulted, as well as the bad, we should remember that the first step in the process was taken on the back of a sea turtle. ∼

Green turtles awaiting shipment by steamer to northern markets. In the 1920s people still caught turtles in the Dry Tortugas Islands and west of the Florida Keys. The turtles were held at Key West, with flippers tied and the weight marked on their plastrons, until they were shipped for slaughter.

CHAPTER FIVE

CONSERVATION

An Uncertain Future

WHEN CHRISTOPHER COLUMBUS FIRST SAW THE CAYMAN ISLANDS he was steering his ship through a sea of turtles. His son Ferdinand wrote, "We sighted two very small low islands full of turtles (as was all the sea thereabouts, so that it seemed to be full of little rocks); that is why these islands were called Las Tortugas." The turtle population appeared to be inexhaustible to the generations of hunters supplying meat to passing ships.

A few centuries later, however, the turtles of Las Tortugas were gone. Hunters sailed for Cuba to feed the continuing demand, but soon the Cuban turtles were extinguished, as well. Most of the populations that remained in the Caribbean shrank because few turtles survived long enough to grow to full size.

As the nineteenth century ended and the practice of stowing turtles in the hulls of boats waned, a new problem arose. The ships brought settlers and created a coastal economy. Villages, towns, and cities sprouted all along the coasts of every continent. The number of mouths to feed ballooned and most of those mouths were within a day's travel of the oceans. The costs of feeding the age of sail were significant reductions in the numbers of sea turtles worldwide and some local extinctions, but the problem was soon to become much larger.

Turtles to Market

At the turn of the twentieth century, sea turtles had a price on their heads. They were a prized source of protein. Along the Central American coast and on most inhabited islands, nesting beaches became more dangerous than ever. Governments organized turtle harvesting efforts, renting sections of their beaches during the nesting season. The *veladors* (those who stay up at night) bought the rights to a section of beach and patrolled it all night. When a green turtle or hawksbill came ashore, the *veladors* flipped it onto its back so it could not escape. The next morning the captured turtles were moved to market.

If the beach was near town or a road the *veladors* dragged the turtles to a waiting cart or wagon. In more remote places they put the turtles in the shade and waited for a boat to come by and pick up the catch. It might be a few days or it could be a week. The wholesaler waited offshore while the *velador* tied a buoyant balsa log to

Previous pages: An olive ridley turtle snagged on a longline set for sharks off Nancite along the Pacific coast of Costa Rica. A paucity of patrol boats makes enforcement impossible in this national park. Longline fishing is one of the greatest threats to sea turtles and may be pushing Pacific leatherbacks to the brink of annihilation.

Opposite: Miskito Indians still rely on green turtles for food, capturing them along the Caribbean coast of Nicaragua. Sailing back to their village the fishermen butcher the turtles and distribute the meat among the residents. However, the catch often exceeds local needs and the extra turtle meat is sold in the marketplace.

each of the turtle's flippers. With the help of family members the turtle was pushed into the breakers and directed toward the ship. The wholesaler sent out small boats of men that grabbed the logs and pulled the turtles aboard. They were then taken to the mothership to join countless other turtles heading to slaughter.

Not all of the harvesters on the beach collected the nesting mothers. *El hueveros* (the egg men) were after the unborn offspring, actually taking the eggs as they came out of the mother. In the morning they looked for nests they missed by poking in the sand with "egg sticks," which were five or six feet long (~2 m) and about as thick as a thumb. The *huevero* would push the stick deep into the sand around a disguised turtle nest, probing until he felt a change in resistance when the stick entered a cavity. The still-fresh eggs were placed in a sack and sold to a wholesaler.

A day or two later the turtles and eggs would be dockside in a port town waiting for shipment to New York, London, Tampa, or Key West. The arrival of steamships and airplanes in the twentieth century fueled demand, as fresh turtle meat and eggs could be delivered to most population centers quickly and efficiently direct from the Caribbean. Considered a delicacy, turtle meat, calipee (cartilage), and eggs became common fare in fine restaurants and in the homes of the wealthy.

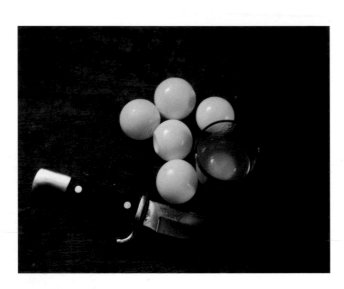

Above: Olive ridley eggs are eaten with a special salsa and washed down with beer by Costa Rican men as an aphrodisiac. The legal market for olive ridley eggs from Ostional provides a cover for illegal collection of eggs elsewhere.

Opposite: Illegal harvesting of eggs from a leatherback turtle in Costa Rica. This villager is typical of many local people who adamantly believe that sea turtle eggs are an inexhaustible resource.

The global impact of the "turtle to market" system was staggering. Sea turtles probably have been taken in small numbers by local tribes for a long time but they can't feed the world. Most species take decades to reach maturity, so any major rise in sea turtle mortality causes them to decline.

Yet even today it is legal in many countries to harvest sea turtles for commercial use. Sea turtle eggs are sold as both a food source and an aphrodisiac. Green turtles are taken for their meat, calipee, and shells. Bones become tools, art objects, jewelry, and fertilizer. Small sea turtles are stuffed and marketed as tourist souvenirs. The skin of olive ridleys is turned into boots, shoes, and handbags. Leatherbacks are rendered for oil. The beautiful shells of hawksbills are sold as wall mounts or turned into jewelry, combs, and eyeglasses.

Many scientists and conservationists think that continuing the legal harvest of the greatly depleted populations of sea turtles is highly irresponsible. Even if one views sea turtles as a resource to be exploited, clearly the animals need a breather. Stopping the legal collection and trade will require a concerted effort, including the production of commercial substitutes, international pressure and embarrassment, education, and enforcement. When it is no longer legal, easy, and profitable, we may well see an end to the turtle trade.

Sea Turtle Conservation: The Ten Best and Ten Worst Countries

Ranking the conservation status of countries is a complex, qualitative affair. With regard to sea turtle conservation, several factors are important when determining how well, or how poorly, a country is doing. The factors considered in the following ranking were: 1) degree of egg harvest, 2) quantity of adult harvest, 3) extinction rate for populations occurring within the country; 4) quality of nesting habitat, 5) quality of foraging habitat, 6) management of unique habitats, 7) presence of endemic sea turtle species and how well the country addresses that issue, 8) controls placed on fisheries including TED requirements and longline fisheries management, 9) impact of fisheries on sea turtle populations. Based on these criteria and surveys of experts, the following list highlights the best and worst countries for sea turtles today.

Ten Best		Ten Worst	
1. Australia	6. French Guiana	1. Japan	6. Indonesia
2. United States of America	7. Oman	2. Taiwan	7. Thailand
3. Costa Rica	8. Suriname	3. Spain	8. Morocco
4. Mexico	9. Greece	4. Korea	9. Cuba
5. Brazil	10. Malaysia	5. Philippines	10. Equatorial Guinea

Unintended Consequences

The turtles whose ancestors did not end up in a ship's hull or on a dining table were far from safe. They suffered directly from the greatly increased demand for fish as the twentieth century progressed. Indeed, the leading cause of sea turtle mortality during the last 50 years has been their unintentional capture by three commercial fisheries: shrimp trawling, gill netting, and longline fishing.

Shrimp trawling was long the single leading cause of death for sea turtles. Estimates suggest that as many as 50,000 loggerheads and 5,000 Kemp's ridleys died each year in shrimp trawls in U.S. waters.

Shrimpers drag large nets behind mid-sized boats. The nets remain underwater for an hour or more depending on the conditions. Unfortunately, sea turtles often frequent the same waters as shrimp, and get swept into the nets. They can hold their breath for about 45 minutes when active, but a forced submergence can lead to problems much more quickly. Depending on how long they are trapped—and how long it has been since their last breath—the turtles may be brought aboard alive, dying, or dead. Even those released alive might well be netted during a subsequent trawl.

A partial solution to this problem situation was found in 1978. Turtle excluder devices, or TEDs, have evolved into a variety of types, but in essence they consist of a trap door that allows a turtle to swim out of a shrimp net just before it enters the rear bag where the shrimp accumulate. Several studies suggest that TEDs reduce turtle mortality by about half and, as an added benefit, reduce the capture of unwanted fish by as much as 50-70 percent. The use of TEDs is now required in many countries

including the United States, but the trawling problem is far from solved. Some countries do not require TEDs or do not enforce the requirement. Fishermen sometimes deactivate the TEDs after they leave port because of their widely held belief that TEDs significantly reduce the shrimp catch.

A second approach to reducing the impact of shrimp trawling is to limit trawl location and timing. Sometimes thousands of turtles are killed in only a few weeks when trawling takes place in areas where turtles are gathering for the nesting season. The damage is so obvious that it is hard to envision credible opposition. However, just as many turtles are killed slowly, one by one, over the entire trawling season. And with tens of thousands of trawlers catching shrimp the numbers add up. Trawl operators, for the most part, are not wealthy and they plead their cases strongly to sympathetic governments. The battles often come down to rhetoric: "Salt of the Sea" versus "Tree Hugger." Yet the damage that trawling has done, and continues to do, to sea turtle populations seriously undermines the conservation of these species.

It is likely that some fishermen have good ideas about how to solve this problem and it would be helpful to draw them into the search for a solution. This approach might also facilitate the process of getting official approval and implementation of any such solutions.

After shrimp trawling, gill nets are probably the second leading cause of sea turtle mortality. These floating nets, made of a mesh designed to capture fish, kill turtles that become entangled. As a fish swims into the net, it slips its head into the mesh but can only go so far. If it tries to back up, the gill cover catches on the mesh. It cannot

Trawling is unselective and catches turtles, fish, and other sea animals in addition to the target harvest, which is often shrimp. A sea turtle caught in a trawl will drown if it cannot get to the surface in about 40 minutes.

go forward and cannot back up. This very effective way of fishing is also rather undiscriminating. A sea turtle that swims into a gill net can easily get its flipper caught in the mesh. Unless the caught turtle is near the surface, or lucky enough to become entangled just before the net is hoisted, it will die.

In the 1980s massive gill nets were set in the open Pacific Ocean far from shore. Hundreds of miles long and hundreds of feet deep, these nets trapped anything of size in their path. Sea turtles were no exception, and as a result there was a precipitous decline in Pacific sea turtle populations. Fortunately, those nets are now banned. However, there are reports of ghost ships and illegal gill-netters still working the high seas. In addition, coastal gill-net fisheries remain in operation and continue to kill vast numbers of turtles. For example, the Chilean gill-net fishery,

which operates near the coast and targets swordfish, played a significant part in the decline of leatherbacks in the eastern Pacific. Even smaller-scale gill-netting by local fishermen has had a tremendous impact on sea turtles in some areas.

Although it is not a complete solution, it is possible to use gill nets without killing sea turtles. To do so, the fishermen have to use small nets and tend them often (at least hourly) to be sure that no sea turtles are entangled. Biologists use this technique to capture sea turtles for research studies. Such an approach might be quite effective in coastal areas and bays. It is expensive to tend nets in this way, but since this is the only ecologically sound way to use these nets the real cost of such fishing is much greater than we now realize. We either pay more for our fish because more labor is involved in catching them or we pay for the effects of cheaper fish with the extinc-

The turtle excluder device (TED) has been a true lifesaver for sea turtles. A specialized gate allows sea turtles and non-target fish to escape from a shrimp trawl. This loggerhead will live another day as a result of this US$300 modification to the gear.

CHAPTER SIX

GREEN TURTLES
The Grass Eaters

T HE GREEN TURTLE, *CHELONIA MYDAS*, IS A MOST MAGNIFICENT SEA turtle. It is everything one would expect to see in such a creature. Large, with a distinguished, inquisitive face, it flies through the water with ease. Its shell is nicely rounded and gives off a brilliant sheen in the water that slowly fades as it dries. Its well-formed flippers have pronounced scales that grade from large to small to large, from the front edge to the rear. The green turtle's flippers appear a bit small for its body size but are quite efficient for propelling it through the oceans.

Yet the green turtle is probably best known for its green fat and muscles, which are delicious in soup or as a steak. Herein lies the root of its problems.

For centuries people around the world have considered the green turtle an ideal food source. Once abundant, easy to catch in water or on land, and easy to store alive for months, the green turtle supported the colonization of the New World. Even conservationist Archie Carr praised the merits of the famous London Alderman's soup, the clear green turtle soup that was the highlight of official banquets until the late twentieth century. Indeed, it was a favorite of Winston Churchill.

Green turtles feed and nest throughout the tropics and into subtropical oceans. They live in the Atlantic Ocean as far north as Massachusetts and as far south as southern Brazil, throughout the Mediterranean Sea, over wide areas of the Pacific Ocean, and in much of the Indian Ocean. Today green turtles migrate extensively to and from their nesting beaches just as they were observed by Columbus and other early explorers. However, individual turtles spend most of their time in small areas feeding on seagrasses and algae. Often a green turtle will have a favorite sleeping location and return to it night after night.

Truly an international animal, green turtles swim and feed in the coastal waters of at least 140 countries. They lay their nests each year in about 80 of those countries. The largest nesting colonies are at Tortuguero on the Caribbean coast of Costa Rica, where about 22,500 females nest each year, and Raine Island on the Great Barrier Reef in Australia with 18,000 nesting females in a season. Other large nesting colonies include Oman (6,000), the Comoros Islands (5,200), the Seychelles (4,900),

Previous pages: Green turtles congregate in large numbers during the breeding and nesting season on the Great Barrier Reef of Australia.

Opposite: A green turtle resting along a reef on the Island of Sipadan off Borneo. The reef is covered with algae, sponges, and other marine life. Green turtles are not aggressive and often allow divers to get very close.

Green turtles are the only sea turtles that eat large amounts of plants. They consume a variety of seagrasses such as *Thalassia* (turtle grass), *Syringodium, Halophila, Posidonia, Halodule,* and *Zostera,* as well as algae such as *Chaetomorpha, Sargassum,* and *Hypnea.* To a lesser extent they also eat other green and red algae and some jellyfish, salps, and sponges. In the East Pacific they seem to eat more animal prey, including mollusks, fish, polychaete worms, and jellyfish. Green turtles grow quite slowly on their seagrass and algae diets, presumably because the food is very low in protein. Fed a diet rich in protein, such as is provided for captive turtles at the Cayman Turtle Farm, they can grow more rapidly and reach maturity in five to seven years.

It takes longer for a green turtle to reach maturity than any other sea turtle—any other turtle, for that matter. In Costa Rica and Florida females mature at 26 or 27 years of age while those in Australia take 30-40 years because there is less food available. The green turtles of Ascension Island mature at about 35 years and Hawaiian green turtles at 30 or older. This is quite old for any vertebrate and simple math dictates that a high percentage of each year's class of turtles must live to maturity if the species is to survive.

Green turtles typically live about 19 years beyond maturity or 45-59 years. Given that a female nests in six or seven of those years and lays about 330 eggs each nesting season, she will produce 1,900-2,300 eggs in her lifetime. Factoring in natural predation, fungus infections of nests, and other embryo failure rates, one can expect each healthy female to produce 1,000-1,900 hatchlings.

Population Relationships: It's in the Genes

Green turtles in the Atlantic can be genetically differentiated from those in the Pacific Ocean. Sometime in the past these populations apparently were able to mix by swimming through a large land gap between North America and South America. However, about three to five million years ago Central America rose up out of the sea and created the Isthmus of Panama, isolating these populations from one another. The water around the southern ends of Africa and South America is apparently too cold for this tropical species.

Consequently, in the Pacific there is a population of green turtles that looks different from other green turtles. These dark-colored turtles range from Baja California, Mexico, south to Peru and west to the Galapagos Islands. They differ in color, size, shell shape, and the number of eggs they lay (fewer, about 65-90 per clutch). They have even been given a different common name, the black turtle, and some scientists consider them a unique species, *Chelonia agassizii,* and not *Chelonia mydas* as all other green turtles are classified. However, modern genetic analyses have demonstrated fairly conclusively that the differences are minor. Still, some scientists believe the morphological data suggests that the black turtle is perhaps best considered a separate subspecies, *Chelonia mydas agassizii.* In general, external color and

Heat, of cour
temperatures ne
temperatures, bu
where winter ten
subtropical lagoo
exposed to the co

To avoid thes
avoid extreme no
subtropical lagoo
appear to absorb
without coming t
gen, thus allowin

Decline of the G
The historic decli
best-documented
lation was only a
other 50-70 perce

Green turtle being cleaned by reef fishes in Hawaii. Many turtles enjoy a symbiotic relationship with cleaning fishes that help keep their shells free of barnacles, algae, sponges and other
creatures that would use the turtle as a mobile reef. In return for the cleaning (and subsequent
improvement in the turtle's drag coefficient), the fish get an easy meal.

to 58 percent today. Similar estimates of egg collection are believed likely for the Atlantic coast of Africa where on some beaches perhaps half the nesting females are killed each year.

About 100 adult green turtles are harvested in Japan each year for local consumption by villages. Between 5,000 and 10,000 green turtles are killed annually in foraging grounds near Baja California, Mexico. Similar numbers are reported from the Miskito Cays, Madagascar, and Oman, the last population apparently near imminent collapse. Amazingly, these numbers pale in comparison to Southeast Asia, where more than 100,000 juvenile and adult green turtles are harvested annually.

Right: Mockingbirds on the Galapagos Islands are larger than their mainland relatives. This allows them to attack and kill green turtle hatchlings after they emerge from the nest. A mockingbird cannot carry a hatchling, so it pecks it open and leaves the hollow shell behind.

Opposite: A hatchling black turtle swimming gracefully off Michoacan, Mexico, is a hopeful sign for the future. Numbering only 850 where there were once 25,000 nesting females, every hatchling is a valuable addition to the population.

Even Costa Rica, with its laws and active protection, suffers significant losses from poaching.

Green turtles are recognized internationally as endangered by the World Conservation Union (IUCN) and are protected in Appendix I of the Convention on International Trade in Endangered Species (CITES) and in Appendices I and II of the Convention on Migratory Species (CMS). They are listed by the United States and by many other nations as protected in national legislation (for example, the U.S. Endangered Species Act). The recently ratified Inter-American Convention for the Protection and Conservation of Sea Turtles should be a great help in improving protection throughout the Americas. Green turtles are also protected by the Memorandum of Understanding on the Conservation and Management of Marine Turtles and their Habitats of the Indian Ocean and South-East Asia (IOSEA), the Memorandum of Understanding on ASEAN Sea Turtle Conservation and Protection, the Memorandum of Agreement on the Turtle Islands Heritage Protected Area (TIHPA), and the Memorandum of Un-

derstanding Concerning Conservation Measures for Marine Turtles of the Atlantic Coast of Africa.

Because of these laws and agreements the harvest of eggs and adults has been slowed and some nesting beaches have been protected. The use of Turtle Excluder Devices has been implemented, especially in the United States, Mexico, and South and Central America. However, the long list of countries where the taking of eggs and adults still occurs, legally and illegally, indicates that much more needs to be done to save the green turtle from extinction. We can all get involved in this process by joining or supporting a reputable conservation organization that strives to keep green turtles from joining the growing list of extinct wildlife.

The Future

Our goal should be to restore green turtles to population levels at which they can fulfill the ecological roles they performed in the past. In all honesty, most beaches will never be restored to prehistoric levels, at least not for centuries to come. We will likely never see sailboats bumping into herds of migrating green turtles, but we must continue to pressure nations to stop legal and illegal taking of green turtles. Perhaps we can even change attitudes by providing alternative foods for a growing and hungry world.

The movement to save green turtles has begun but it is not yet up to speed. There are many ways to proceed and each needs a champion. Restoring nesting turtles to historic locations like the Dry Tortugas and the Cayman Islands, where they once abounded, would take a lifetime and perhaps more, but what a reward. Fighting for full protection of nesting beaches country by country is among the worthiest of goals. Another way to make a difference is by opposing residential and commercial development of green turtle nesting beaches. And we are still waiting for someone to devise a way to eliminate or greatly reduce the incidence of fibropapillomatosis in green turtle populations.

The commercial fishing industry will adopt techniques that are harmless to turtles only if governments hear louder calls from their citizens than from those who fish. Raise your voice and demand responsible fishing everywhere. Some of us will aid green turtles directly, others through donations, by writing letters, or by making phone calls.

The future for green turtles is uncertain. If nothing changes they will likely become extinct in most of the world during this century, holding on in a few places where protection is strong. If we act, however, they may soon reach the bottom of their decline and begin to climb. If so it will result from practical steps taken one beach, one country, one law, one village at a time. And it will have been worth it. ⌒

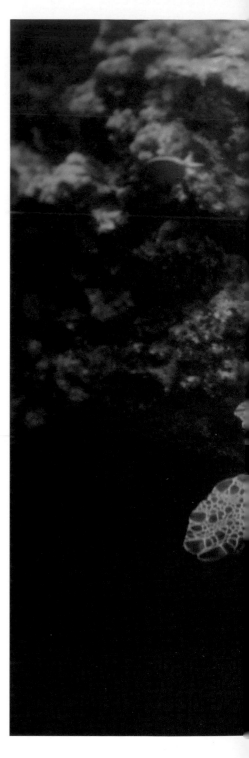

Green turtle drifts over the reef at Sipadan Island in the Celebes Sea.

CHAPTER SEVEN

HAWKSBILLS
The Sponge Eaters

WITH A BEAK LIKE A HAWK, SCUTES MORE LOVELY THAN IVORY, AND a diet of glass, the hawksbill, *Eretmochelys imbricata*, is not only the most beautiful turtle but also one of the most unusual. Egyptians, Romans, Chinese, Europeans, Arabs, American Indians, and Japanese have made jewelry out of the richly patterned scutes that cover its shell. Their scutes are translucent amber with rays of yellow, white, red, black, brown, and gold throughout. This tortoiseshell, or *bekko* in Japanese, is used as a colored inlay in fine antique furniture and to make combs, fans, and various ornamental objects. In my youth a grandmother's dresser top was sure to have tortoiseshell hair clasps and combs, but now these items are more likely to be found in museums. While the tortoiseshell objects are certainly more beautiful than the plastic replacements available today, that beauty is the reason the hawksbill has been hunted nearly to extinction.

Today the hawksbill populations are greatly diminished. Once an animal that nested in colonies, the hawksbill is currently often a lone nester. Once common on coral reefs and a central player in maintaining the structure and function of coral reef ecosystems around the world, today the hawksbill is a rare inhabitant.

But the tide has turned in some sense for the hawksbill. Trade in tortoiseshell is now banned by international treaty, greatly reducing the marketplace for tortoiseshell hair clips. Perhaps as ancestral combs are passed to the next generation it will be with the knowledge that hawksbill populations are growing.

When a hawksbill swims around a coral head and into a ray of sunlight, the resulting flash of color is a near heart-stopping experience. Those lucky enough to see this turtle while snorkeling or scuba diving in the Caribbean, Red Sea, or Seychelles understand immediately why the turtles are better left in the water than cut up for combs and brush handles. Just as important is the fact that live hawksbills are estimated to be worth $30,000 each to a local resort economy.

The hawksbill's unique ecological role is the result of a diet that is often rich in sponges. Dr. Anne Meylan of the Florida Marine Research Institute discovered this fact during her doctoral study of the hawksbill's feeding ecology. She found that sponges account for 95.3 percent of the Caribbean hawksbill's stomach contents. Previously it was assumed that Caribbean hawksbills ate a much more varied diet of plants and animals.

Previous pages: A hawksbill turtle swims through the water with two remoras attached to its plastron. Sea turtles essentially live in a zero-gravity environment because they are neutrally buoyant and able to cut through the water with ease.

Opposite: Hawksbills are the architects of coral reefs. By eating a diet of sponges they control the community structure of the reef and promote the growth of corals.

Meylan's discovery, published in *Science*, one of the world's leading scientific journals, made her famous but cost her a hand. We think of a sponge as a soft material that soaks up soap and water and is useful for washing dishes and cars. A living sponge, however, is one of the most primitive multicellular animals on earth. Lacking a backbone, it has instead a skeleton made of spicules, or spines, composed of calcium, silica, or protein depending on the type of sponge. These spicules are shaped like needles and are curved or straight with hooks and knobs. Hawksbills prefer sponges that have silica spines in their body wall.

While conducting fieldwork, Meylan was handling a sponge and some of its glass spicules stuck in her fingers, some breaking off beneath the skin. In those days we didn't wear thick gloves when we did such work. And who would have thought that a hawksbill's stomach was filled with small glass shards? Years later Meylan noticed a small tumor in one of her fingers. Doctors suspected that it had formed around one of those little glass spicules, just as a tumor forms around an asbestos fiber in someone's lung. When other treatments failed, they were forced to remove the finger. But by then the tumor had spread to her hand, so that had to go as well. Fortunately that was the end of her troubles. Unflappable and committed to sea turtle conservation, she remains one of the world's experts on hawksbills and an active researcher. Now she just does it all left-handed.

Distribution

Hawksbills are distributed worldwide in coastal waters within tropical and some subtropical seas. They migrate hundreds of miles between their feeding grounds and nesting beaches even though seemingly suitable nesting beaches are within sight of the feeding grounds. Even hawksbill researchers cannot explain this behavior. Like green turtles, hawksbills forage mostly in small areas over coral reefs, rock outcroppings, and sea grass beds in mangrove-lined bays and estuaries. They will also inhabit lagoons with mud bottoms as well as narrow creeks and passes. Also like green turtles, they are known to return to the same sleeping spot night after night.

Hawksbills live in the waters of at least 82 nations and nest on the beaches of 60 of those countries. The largest populations live in the Caribbean Sea, the Seychelles, Indonesia, and Australia. There are no hawksbills in the Mediterranean Sea. Although population estimates are plagued by a lack of detailed information for many areas, the best estimates suggest that 20,000-26,000 female hawksbills dig nests each year. Because they nest once every 3 years (on average) there are probably 60,000-78,000 adult female hawksbills in the world today. There are only a few places where 1,000 or more females nest each year. The largest nesting populations appear to be in Australia, where 2,000 turtles nest on the northwest coast and 6,000-8,000 on the Great Barrier Reef. The Mexican states of Campeche, Yucatán, and Quintana Roo combined host some 2,800 nesting hawksbills annually. Indonesia hosts about 2,000 nesting hawksbills on

Above: Anne Meylan holding a hawksbill in the Miskito Cays in 1978. Archie Carr had just bought the turtle from local fishermen and after this picture Anne set it free.

Opposite: A hawksbill covers her nest after laying eggs on Gulisaan Island in Malaysia. Hawksbills are fast nesters and can complete the process in less than 45 minutes.

its many mainland and island beaches each year, and about 1,000 females nest in the Seychelles. Many of the nesting beaches that remain are visited only by scattered individuals that may be remnants of nearly decimated populations.

Very few hawksbills inhabit U.S. waters, although those that do are fairly well protected by government agencies. Some 120-200 hawksbills nest on Mona Island off Puerto Rico and another 35 or so nest in the Buck Island Reef National Monument at St. Croix, U.S. Virgin Islands. Sadly, only one or two hawksbills nest in Florida each year.

Today hawksbills represent less than 10 percent of the population present a century ago, which in turn was probably less than 5 percent of the turtles pres-

ent centuries before. The decline over the last 100 years is not surprising when figures for the *bekko* trade are examined. Indonesia exported more than 700,000 juvenile and adult hawksbills as stuffed curios and tortoiseshell between 1970 and 1986. In the Seychelles, 50-75 percent of all nesting females were killed during the nesting seasons of 1980, 1981, and 1982. Madagascar exported tortoiseshell from more than 5,400 hawksbills per year in the nineteenth century, but by 1950 it had slowed to 1,300 per year. By 1973 there were too few hawksbills to support a trade.

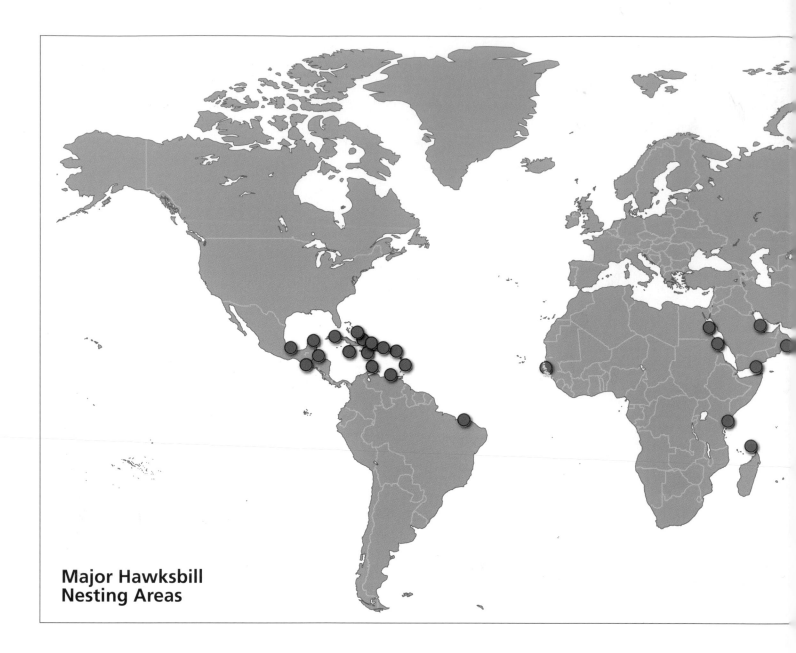

Major Hawksbill Nesting Areas

Exploitation in Panama began in 1815 on Chiriquí Beach, which was famous for its huge nesting colony. More than a century later, *veladors* each caught 30-50 hawksbills a night per mile (1.6 km) at the peak of the nesting seasons of the early 1950s. In 1980 and 1981 only 17 and 13 hawksbill tracks, respectively, were found on the entire 17.5-mile-long (28 km) nesting beach. By 1990 only one nest was seen on the beach. Today, restoration efforts are in place and as many as several hundred nests have been documented in recent years. It seems the tortoise shell ban happened just in time to save this population.

Hawksbill tracks are quite distinct because of the alternating style of locomotion. In Tortuguero, Costa Rica, I once came across a set of hawksbill tracks leading from the water and followed them up the beach. They led toward the jungle but suddenly stopped halfway up the beach. It was as if the turtle simply flew away. There was no

Population Estimates for Areas with Major Hawksbill Colonies

Location	Number of Nesting Females per Year
Caribbean Sea	**5,000- 6,000**
Antigua/Barbuda	150
Barbados	50-60
Belize	40-50
Cuba	500-1000
Dominican Republic	310
Guatemala	100-200
Jamaica	200-275
Martinique	80-125
Mexico	2,800
Puerto Rica	210
Turks and Caicos	200-275
U.S. Virgin Islands	130
Venezuela	50-500
Northern South America	**600**
Brazil	400-500
Eastern Atlantic Ocean	**200-400**
Guinea Bissau	200
Indian Ocean	**6,000-7,000**
Andaman and Nicobar Islands (India)	250
Australia (North West Shelf)	2000
British Indian Ocean Territory (Chagos Archipelago and southern Maldives)	300-700
Burma	30
East Africa	100
Egypt	500
Maldives	300
Oman	600-800
Saudia Arabia	160
Seychelles	1,000
Sudan	350
Yemen	500
Pacific Ocean	**10,000**
Australia (Great Barrier Reef to Arnhem Land)	6,000-8,000
Indonesia	800-2,000
Malaysia	100-500
Palau	20-50
Papua New Guinea	<100
Philippines	100-500
Solomon Islands	<500
Thailand	<100

nest and no return track. For a moment I stood there bewildered. Then I noticed a human's bare footprints and realized someone had picked up the hawksbill (a small species) and carried it off into the jungle. Since we had been patrolling the beach every half hour I knew he could not have gone far. I moved quickly into the vegetation, following the erratic pattern of prints, but to no avail. They disappeared into the jungle without a trace and I was unable to aid the doomed turtle. Knowing that I had been within 10-30 minutes of saving that one turtle, the image of those tracks disappearing into the jungle stayed with me for months.

The World Conservation Union lists hawksbills as "critically endangered" on its Red List of Endangered Species. Yet in many places people still kill and eat hawksbills and dig up their eggs. In fact, almost every nesting female in the Philippines ends up on a supper table or in a souvenir shop. However, the future for this species looks brighter than it has in decades. With a continued ban on international trade, more ecotourism (thus increasing the value of every live hawksbill to locals), and increased education and enforcement, there is hope that the numbers of hawksbills will begin to rebound in many areas.

Nesting

In the years a female hawksbill nests, she lays 3-5 clutches of eggs at an interval of 13-16 days between nests. Each nest contains an average of 130 eggs. With each egg weighing about 1 ounce (28 g) the average female will lay about 8 pounds of eggs each time she nests. It takes 60-90 minutes for a hawksbill to complete the nesting process, which usually takes place at night and often occurs under trees and vegetation, especially in the Caribbean and Australia.

Hawksbill tracks are easy to distinguish from those of green turtles owing to the alternate use of flippers as they pull themselves up the beach. They move quickly to their nest site and are much faster nesters than green turtles. Sometimes they almost seem to "run" back to the water after completing their nest and laying their eggs. Perhaps the many generations of hunting for hawksbills have resulted in natural selection for fast-nesting turtles; only the fastest nesters survive to reproduce again.

With the nests dug, the eggs laid, and their mother trotting back to sea, the offspring are left to their fates. The pivotal temperature for sex determination is approximately 85°F (29.4°C). A nest at 86.5°F (30.3°C) produces 100 percent female hatchlings. If humans do not disturb the nest, hatching success can run from 79-92 percent on some beaches in some years. On other beaches, predators can take many if not most of the nests. Caribbean hurricanes can wreak havoc in some years.

When hatchlings break free of the egg, they weigh about 0.5 oz (14 g). After emerging from the sand at night, the neonates race to the sea and swim for several days. They spend their first 1-3 years at sea, often living in floating rafts of *Sargassum* where they remain motionless for long periods of time. Young hawksbills are rarely seen so we do not know much about this phase of their life, but it appears that they are omnivores. The stomach contents of young turtles that washed ashore dead in Florida have included *Sargassum*, other algae (*Syringodium* and *Microdictyon*), goose barnacles, fish eggs, tunicates (marine animals, such as sea squirts and salps, having a cylindrical or globular body with a tough outer covering, or tunic), and crabs.

When hawksbills are 1-3 years old and reach 8-10 inches (20-25 cm) they move onto reefs or into bays in the Caribbean. In the Indo-Pacific and Australia they move closer to shore when they are larger, about 15 inches (38 cm) long. Once on their feeding grounds, Caribbean hawksbills may continue to eat a mixed diet of invertebrates, fish, algae, tubeworms, clams, and snails. But they quickly begin to eat sponges. As they grow larger they eat more and more sponges, soon eating sponges almost exclusively. In the Indo-Pacific they continue to eat a mixed diet that includes

sponges, demonstrating a behavioral difference within the species between the two oceans.

Changing their diet to sponges, however, is not easy. First the turtles have to find out where the sponges live and which ones to eat. This is a tricky business. Eating the sharp, glass-like spicules and their toxic chemical compounds would kill a human. Even hawksbills require some time for their physiology and their digestive tracts to adjust to this diet. Once they do, however, hawksbills are not harmed by either the physical or chemical deterrents. In so doing, the turtles may find that they have a nearly exclusive food source, with little competition from other animals.

Perhaps most surprising is the fact that Caribbean hawksbills feed exclusively on only a few kinds of sponges, such as those in the genera *Chondrilla*, *Ancorina*, *Geodia*, *Placospongia*, and *Suberites*. Hawksbills appear to avoid sponges with skeletons made of the protein spongin or of calcium carbonate.

In some ways it is amazing that it took scientists so long to figure out what the larger hawksbills eat. Local residents of coral reefs knew that hawksbill meat was poisonous, but we scientists didn't ask "Why?" until Anne Meylan carried out her famous diet study some 25 years ago. What other ocean mysteries can be solved by inquisitive and clever young scientists who ask just the right question?

A hawksbill turtle feeding on a large sponge. The hawksbill is known for its hawk-like beak, which is made of keratin. The pointed snout allows the turtle to bite off pieces of large sponges and to poke its face into small crevices in reefs to pick out smaller sponges.

In the Genes

Studies of sea turtle DNA indicate that hawksbills are most closely related to the loggerhead and ridley turtles, which are essentially carnivorous species. It is likely that hawksbills were also carnivores in the past and later developed the ability to eat sponges. They evolved a sharp pointy beak that allows them to pick sponges out of cracks and crevices in coral reefs and bite off pieces of sponge from among mangrove roots. In the months leading up to nesting, females are able to bite off pieces of coral, apparently to get extra calcium for making their eggshells.

Once hawksbills settle onto a reef they grow faster than they did at sea. When they are about 20-24 inches (50-61 cm) long their growth slows until females become sexually mature at about 31 inches (78 cm) and males at about 27 inches (69 cm; in the Caribbean at least). We do not know how long it takes for a hawksbill to mature, but based on the age of maturity for the somewhat smaller Kemp's ridley and larger loggerhead it is probably 20-25 years. Indo-Pacific hawksbills, like green turtles and loggerheads, take longer—about 30 years or more—to reach sexual maturity than hawksbills in the Caribbean.

Architects of the Coral Reef

Hawksbills are not only a beautiful addition to coral reefs but also have an essential role in maintaining the health of those reefs. The physical and biological structures of many coral reefs are directly dependent upon the feeding habits of hawksbills. Reef corals are animals related to sea anemones but different in that they secrete a calcium carbonate skeleton within which the animal lives. In addition, corals share their homes with symbiotic algae, tiny plants that live within the tissues of the coral and provide oxygen and perhaps carbohydrates to the animal.

Sometimes, however, there can be too much of a good thing. So many non-coral animals and plants end up growing on the coral reefs that they can choke out the coral or block the sunlight upon which the symbiotic coral algae depend. Without the algae the coral animals eventually die. This is where the hawksbill comes into the picture.

Sponges are one of the main competitors for space on a reef. Left alone, they can outgrow the coral and cover it up, eventually killing the reef. An average hawksbill adult weighs 154 pounds (70 kg) and eats an average of 1200 pounds (544 kg) of sponges a year. Two hundred years ago there were some 540,000 adult hawksbills in the Caribbean eating 648 million pounds (294 million kg) of sponges each year. They were likely the architects of the reefs, controlling the numbers and size of sponges present and limiting competition between sponges and corals.

The most common sponge in the hawksbill diet is *Chondrilla nucula*. Because there are so few hawksbills left to eat it, this sponge now dominates many reefs, taking up about 12 percent of the reefs in Puerto Rico and accounting for 70 percent of the coral overgrowths on Florida reefs. We know that the decline of coral reefs has accelerated during the last century, especially the last 50 years. Debates continue

This hawksbill is eating small sponges that live among organ pipe coral on a reef along Papua New Guinea. The stinging coral burns the turtle's face as it reaches down into the crevices to bite off sponges.

composition to a nesting population from a particular area. There is a lot more work to be done to fully understand the interaction of hawksbill populations in the Caribbean. However, one thing is clear: Turtles swimming on Cuban reefs or those of any other nation are not the property of that nation alone. They are a shared gift that belongs to all of the peoples of the Caribbean.

The same story is beginning to unfold in the Pacific Ocean as well. Hawksbills that nest in Indonesia and Okinawa feed as far away as the Solomon Islands and the Philippines. Australian hawksbills feed in Papua New Guinea, the Solomon Islands, and other distant places. It is clear that hawksbills feeding anywhere in the world actually hatched on many different nesting beaches. Thus harvesting hawksbills on their feeding grounds will affect the populations on nesting beaches both nearby and far away. The maternal lines of turtle populations from different nesting colonies are isolated from one another, so each nesting population has a unique genetic profile. If a population is wiped out by excessive hunting, as has happened at some nesting colonies in Cuba, it is unlikely that other turtles will recolonize those nesting sites any time soon.

The Future

The sharp decline in hawksbill populations during the latter half of the twentieth century, following 300 years of reductions, makes for a somber story. Even those who deal with the statistics on a day-to-day basis find them depressing. Turtles are still being slaughtered in Indonesia and most eggs are taken in Malaysia.

We can take heart, however, in some recent changes that are improving prospects for the hawksbills that remain. Most nations have enacted laws to protect the turtles and their eggs. The legal international trade in tortoiseshell has all but ceased. Populations are increasing in Mexico and at protected beaches like Jumby Bay in Antigua. More and more nations are requiring TEDs to be used by their shrimp trawlers.

With our increased understanding of hawksbill behavior, migration, and population dynamics, we are better able to establish management and protection plans for their recovery. It is clear that the species is in a precarious position and that all eggs on all beaches should be protected. Killing adults, even for local consumption, must stop. All nations must recognize that hawksbills are shared resources and not the province of any one nation. One of our most difficult tasks is to develop local conservation plans for the many places where hawksbills live and nest. When local communities learn that a live hawksbill is worth more than a dead shell, they may begin to protect their own turtles. Like politics, all conservation is local. We have the knowledge to develop local, national, and international conservation measures to recover hawksbill populations in many areas of the world. Who will step up to the plate and take a swing at modern conservation to save turtles and help communities at the same time? ～

Hawksbills, such as this one moving along the Florida coast, undertake long migrations from their feeding grounds to the nesting beaches. Protection is needed at both ends of their migratory journey if the species is to survive.

CHAPTER EIGHT

OLIVE RIDLEYS

A Phenomenon of Nature

THE OLIVE RIDLEY TURTLE IS A BEAUTIFUL DANCER. ON CERTAIN CLOUDY, moonless nights you can hear the female's dance as she covers her nest of eggs. Long before you see her a series of thumping sounds can be heard in the night air. Then you spot a little (by leatherback standards) turtle bouncing on the sand, pounding it flat and hard with her plastron. Propped up on her front flippers and hind legs, she alternately bounces her plastron side to side in a little dance that seals the site where she has just deposited her offspring. It strikes one as unusual and unexpected: Look at the silly turtle dancing on the sand.

Most sea turtles cover their nests with their hind flippers and throw sand with their front flippers. These species have enough mass and strength to compact the sand with their rear flippers and tail. The relatively light olive ridley, however, must pound the sand down like the head of a hammer, dancing her way around the place where she has just put the next generation.

Olive ridleys do something else that is phenomenal. When it is time to nest, most olive ridleys come ashore in groups of many thousands. The Spanish call these arrivals *arribadas* and they are arguably the greatest display of animal reproduction in the world. Imagine 50,000 olive ridleys laying 110 eggs each on 1 mile (1.6 km) of beach, coming ashore in wave after wave over a 3-day span. That's a total of 5.5 million eggs weighing 1.4 ounces (40 g) each or 485,000 pounds (220,000 kg) of eggs.

Another striking thing about olive ridleys is that they nest in the daytime as well as at night. Some years ago my colleague Ed Standora and I demonstrated that large sea turtles could not nest during the heat of the day because they became overheated. Like Kemp's ridleys, which nest mainly in the daytime, and flatbacks, which often nest during the day, olive ridleys are able to accomplish this feat because of several adaptations. First, they avoid nesting during the hottest part of the year and during the hottest part of the day. Occasionally, you can find turtles of many species, even leatherbacks, nesting during the morning. Presumably they came out late at night on a morning high tide and were greeted by the sun. Ridleys, however, purposefully nest during the morning and late afternoon.

Intuitively we might think that a small sea turtle would heat up faster than a large one, just as a small block of ice melts faster than a large one. However, olive ridleys are

Previous pages: A small olive ridley *arribada*, or mass nesting, at Ostional on the Pacific coast of Costa Rica. At the height of an *arribada* the beach is literally a blanket of turtles, digging nests, covering them up, and walking to and from the ocean. Turtles come and go for two or three days at a time, resulting in thousands—or perhaps millions—of eggs in the sand.

Opposite: Olive ridleys clambering up the beach to dig nests and lay eggs at Ostional, Costa Rica. Ridleys plod along with an alternating gait that leaves a characteristic track. During an *arribada* they seem unconcerned by the presence of humans.

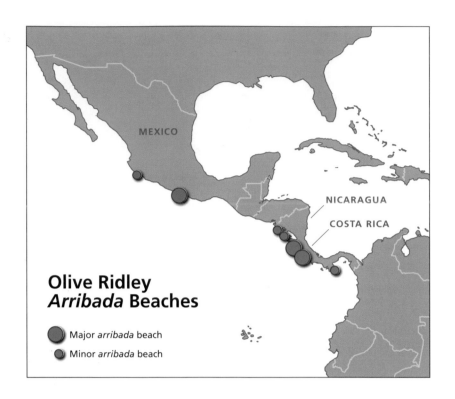

Olive Ridley
Arribada Beaches

⬤ Major *arribada* beach

⬤ Minor *arribada* beach

lighter in color than the larger green, loggerhead, and leatherback turtles, so they absorb less solar radiation. Their smaller size also allows them to cool more effectively from the convective action of wind. *Arribadas* are often associated with windy and cloudy days, so for added protection they apparently select advantageous weather conditions, thus reducing the heat load on the turtles and increasing heat loss by convection. In addition, they are fast nesters. Although most sea turtles take 60-90 minutes to nest (green turtles can take 2-3 hours), ridleys complete the entire nesting process in less than an hour.

Distribution

Olive ridleys thrive in the tropical to warm-temperate Pacific and Indian Oceans, but are uncommon in the western Pacific and eastern Indian Ocean. They also inhabit the Atlantic Ocean along the west coast of Africa and northeastern coast of South America. Although olive ridleys are seen as far north as Alaska and as far south as New Zealand and Chile, their infrequent visits occur only during unusually warm periods. While a few olive ridleys are present in the West Indies, they do not occur in the Gulf of Mexico or in most of the North Atlantic Ocean.

About 3-5 million years ago, when the Isthmus of Panama formed and closed off the Atlantic from the Pacific, olive ridleys and Kemp's ridleys began traveling down different evolutionary paths. We believe that at the time only the ancestors of the Kemp's ridley sea turtle lived in the western Atlantic. In the 1990s, geneticist Brian Bowen conducted studies that suggested olive ridleys spread from the eastern Pacific to the western Pacific and Indian Oceans. Olive ridleys in the eastern Pacific then died out apparently, perhaps owing to climate change or loss of nesting beaches. When climatic

Countries with Populations of Solitary Nesting Olive Ridley Turtles Producing 100 to 2,000 Nests a Year

Region	Nation	Beaches
Western Atlantic	French Guiana	Malmonoury Beach
	Brazil	Sergipe
West Africa	Ghana	
	Gabon	
	Angola	
	Guinea Bissau	Bijagos Archipelago
	São Tomé	
	Equatorial Guinea	Bioko Island
	Togo	
	Benin	
	Cameroon	
East Pacific	Guatemala	Hawaii
	Honduras	Punta Ratón
	Costa Rica	Guanacaste, Playa Hermosa and Osa Peninsula
	El Salvador	Toluca and San Diego
	Mexico	El Quelite, Sinoloa; Chacala, Nayarit; Maruata, Colima; Mexiquillo, Michoacan, Cuajinicuilapa, Guerrero; and 42 more

Region	Nation	Beaches
Western Indian Ocean	South Africa	
	Mozambique	
	Kenya	
	Tanzania	
	Oman	
Central Indian Ocean	Pakistan	
	India	Saurashtra Peninsula and Bhaidar Island; In Gujarat; Tamil Nadu; Andhra Pradesh; South Andaman Island; Great Nicobar Island
	Sri Lanka	
	Bangladesh	
Eastern Indian Ocean	Burma	
	Thailand	
Western Pacific Ocean	Australia	Crab Island
	Brunei	
	Malaysia	
	Indonesia	
	Vietnam	

Major and Minor Olive Ridley *Arribada* Beaches

Beach	Nation	Number of Turtles
Major		
Ostional	Costa Rica	500,000
Nancite	Costa Rica	100,000
La Escobilla	Mexico	450,000
Gahirmatha	India	135,000
Eilanti	Suriname	Disappeared
Mismaloya	Mexico	Disappeared
Piedra de Tlacoyunque	Mexico	Disappeared
Chacahua	Mexico	Disappeared

Beach	Nation	Number of Turtles
Minor		
La Flor	Nicaragua	10,000-20,000
Chacocente	Nicaragua	2,000-5,000
Isla Cañas	Panama	2,000
Ixtapilla	Mexico	2,000
Barunei River	India	2,000-10,000
Masachapa	Nicaragua	Disappeared
Pochomil	Nicaragua	Disappeared
South Andaman Island	India	Disappeared

conditions improved about 200,000 years ago, olive ridleys expanded out of the Indian and western Pacific Oceans and recolonized the eastern Pacific. At that time, the genetic lineages suggest, they also moved into the southern Atlantic Ocean.

Olive ridleys can be seen in coastal waters, but scientists have reported observing them far out to sea in drift lines of seaweed (*Sargassum* algae) and floating debris along convergence zones between warm and cold water regions.

While solitary nesting occurs on the beaches of 32 countries, *arribadas* are found in only a handful. One of the *arribada* beaches, Playa Nancite in Costa Rica, is protected as part of Santa Rosa National Park. Another Costa Rican beach, Playa Ostional, is a national wildlife refuge, but it is also the site of a village of 90 houses. Ostional's people collect olive ridley eggs in a legal harvest. Little information is available about the effect of this harvest on the turtle population.

About two months after an *arribada* the sand releases thousands of hatchling olive ridleys. They race to the sea past egg shells and dead eggs that were thrown out of nests by turtles nesting in the last part of the previous *arribada*.

Nomads of the Ocean

Olive ridleys embark upon unique and complex migrations after the reproductive season. They are true oceanic nomads and swim hundreds or thousands of miles over vast reaches of the ocean to their oceanic feeding grounds. They move away from the beach as individuals take varied paths. For example, some Nancite turtles move along the coast from Costa Rica to Ecuador while others move out from shore to the north, west, and south. They do not move in direct migratory corridors but rather seem to wander over vast stretches of the ocean, possibly using water temperature more than any other environmental cue.

It appears that the olive ridleys are seeking oceanic features that move, such as thermal fronts and the edges of cool water masses. The location of these features may not be predictable to the turtles, so they have to wander and search vast areas of the ocean for appropriate places to feed.

Like other sea turtles, olive ridleys probably use the position of the sun as well as magnetic signals to locate general areas in the ocean. The large amount of magnetite in their brains may help them detect the earth's magnetic field. Some years ago the presence of an internal compass became clear to me in a strange way. Jack Rudloe, an expert on loggerheads, was visiting us at Playa Langosta on the Pacific coast of Costa Rica. One of our students found a dead olive ridley hatchling on the beach in the morning and brought it back to the lab to show us. Jack took the opportunity to demonstrate the magnetic properties of the ridley brain in a very convincing fashion. He placed the rid ley hatchling on a small sponge in a large pan of water and was able to pull it and spin it around by placing an ordinary magnet near its head. If he placed the magnet near the tail, the turtle slowly spun around and moved toward the magnet headfirst.

Olive ridleys probably remember good feeding grounds from the past and move in that general direction. They also wander and move with the shifting water masses as the ocean changes over the course of the year. In El Niño years there is little mixing of cold and warm water in the tropical Pacific and thus the ocean is not very productive. Olive ridleys have a hard time finding food and do not return to nest for two or more years. However, in La Niña years, when cool water upwellings and cold-water masses are more common, the ocean is very productive and food is plentiful. As a result the turtles may nest in consecutive years.

Breeding Behavior

Most adult males and females migrate toward the coast and assemble at breeding grounds off the *arribada* beaches. Some males, however, choose not to migrate and stake their claim on the oceanic feeding grounds. There they intercept females en route to the breeding grounds and nesting beaches. After mating, *arribada* females remain close to shore for several weeks or even a few months and periodically join *arribadas* to lay their eggs. As a result there are vast aggregations of ridleys in the ocean near nesting beaches for several months during the main nesting season.

Solitary nesting females go ashore alone throughout most of the species' range and lay two clutches a season at 14-day intervals. They may use geographically distant beaches within the same season. *Arribada* nesting females have strong nest site fidelity within and between seasons.

Dr. Pam Plotkin tracked the migrations of olive ridleys during *arribadas* by attaching satellite radio transmitters to their carapaces while they were on the nesting beach at Nancite, Costa Rica. She found that females laid two clutches at approximately 28-day intervals, sometimes delaying nesting for 6-8 weeks when environmental conditions were unfavorable.

In Costa Rica, olive ridleys nest in every month of the year but peak nesting occurs in the latter half of the rainy season from August to December. *Arribadas* in the dry season, from January to April, and in the beginning of the rainy season tend to be smaller and last one or two nights. Most *arribadas* start on a windy night within a few days of the first or last quarter moon. Once mating and nesting is completed olive ridleys quickly migrate to their oceanic feeding grounds.

A Bountiful Feast

Predators swarm *arribada* beaches during the nesting season. It is a feast that may provide sustenance for animals from miles around. The beach is literally bursting with millions of eggs. Even before the *arribada* starts, animals begin to assemble at the beach. Vultures sit on tree branches and coatis line the edge of the forest. When we were at Playa Nancite studying the conditions on the beach, coyotes wandered through our campsite at night, as did raccoons and other animals. They seemed to know that the turtles were coming.

Once the turtles start to come ashore the feast begins. Coyotes kill and eat adult females on Playa Naranjo, Costa Rica. Jaguars did the same until they too were reduced to a fraction of what they once were. At Playa Nancite vultures and coatis seem to have worked out a cooperative system for digging up nests. Coatis are good diggers but do not see very well. Vultures have excellent eyesight but cannot dig up a nest. So the coatis dig up the nests and as eggs are flung out the vultures grab some. If a large predator like a coyote, jaguar, or human gets close, the vultures squawk and fly off. That alerts the coati, which then runs away. A bonus for the vulture is that when the coati returns it may dig up a different nest and leave the open one for the vulture to pick over.

As the *arribada* progresses and turtles try to dig their nests they inadvertently destroy many of the nests from previous days. While all this is going on raccoons, coyotes, and other predators join the coatis and vultures in feasting upon the eggs

Above: Dogs and pigs from the village freely dig up nests on Playa Ostional, Costa Rica. Nests are shallow, and broken eggs from an *arribada* often lead animals directly to the feast. On natural beaches eggs are an important resource for coatis, raccoons and other wildlife.

Opposite: Vultures readily eat single hatchlings on Playa Nancite. When thousands of hatchlings come out of the sand at once they can swamp the predators.

that are inevitably strewn about the beach. By the time the *arribada* is finished, the beach from a distance looks like it is covered with giant scattered snowflakes, but up close it smells like a garbage pile because of all the rotting eggs. Predators continue to feast on eggs for the entire incubation period, and then when the hatchlings emerge the animals gorge themselves on the little turtles.

This is one of nature's great examples of the movement of nutrients. The ridleys bring literally tons of nutrients to the beach, where most are redistributed to the animals of the nearby forest. Moving resources from the ocean to the land keeps the dry forest ecosystem alive and healthy.

The Hand of Man

Nowhere in the sea turtle world has the effect of humans been as obvious as on the *arribada* nesting beaches of olive ridley turtles. Nowhere has the slaughter at sea been as great as in the flotillas of olive ridleys that congregate offshore waiting for a chance to mate or lay their eggs. Industrial turtle harvesting began in Mexico in the 1960s, and the death toll was horrendous. Between 1965 and 1970, 2 million olive ridleys were taken from the waters off of Mexican beaches or slaughtered when they came ashore to nest. During the late 1970s in Ecuador as many as 148,000 olive ridleys were killed each year.

CHAPTER NINE

KEMP'S RIDLEYS
The Rarest of All

LOGGERHEADS

A Crushing Jaw

THE LOGGERHEAD TURTLE, *CARETTA CARETTA*, IS A BRUISER OF A TURTLE with a huge head and large crushing jaws. The size of the head-jaw combination allows it to feed on large shellfish and contributes to its reputation as one "tough customer." Stories abound that testify to the nasty disposition of loggerheads. Unlike female green turtles and leatherbacks, loggerheads will bite if you disturb them while they are nesting.

In his book *Handbook of Turtles,* Archie Carr relates a story from a 1905 issue of the *New York Herald* in which five men set out in a rowboat to catch a 610-pound (277 kg) loggerhead in Long Island Sound off Connecticut. Several hours later they limped back into port after trying to spear the turtle, fighting it with their oars, and struggling to stay afloat after the battle. The turtle nearly overturned the boat with its flippers, chewed the oars into stumps, and tore a gash in one of the men's arms. The turtle then sank into the sea and swam away. The men resolved never again to hunt loggerheads. If true, this particular turtle, at least, knew how to defend itself and lived up to the loggerhead reputation.

Among sea turtle lovers the loggerhead has long suffered being "the other turtle." Its carapace, often covered with living things like an old log at sea, is not as beautiful as that of the hawksbill. Its flesh tastes fishy compared to that of the green turtle and the loggerhead is just not as mysterious as the flatback. It is not as large as the leatherback and cannot swim as fast or dive as deep. And loggerheads do not nest in huge *arribadas.*

So, to many people, loggerheads have traditionally been the turtles "next door," ordinary sea turtles that could be seen nesting on the coast of the southeastern United States and on resort beaches in Greece. Only recently has the loggerhead begun to achieve status as the best-known and most-studied sea turtle. It has become a symbol of the conflict between fishing activities and beach development and of sea turtle survival in general.

Despite its commoner status, in reality the loggerhead is a noble species. The female's carapace is reddish-brown blended with olive and its plastron cream to yellow. Her skin is reddish-brown and yellow with tinges of orange. The male's skin contains more brown tones. Hatchlings' carapaces are dark brown or reddish-brown and their plastrons are creamy white to reddish-brown or dark brown.

Previous pages: A loggerhead hatchling goes after food in a *Sargassum* patch in the Atlantic Ocean. Floating seaweed makes an ideal home for little turtles. The rafts of plants provide food, a place to rest, and camouflage from large fish and birds.

Opposite: Loggerhead turtle cruises over the Great Barrier Reef in Australia. The large jaw allows loggerheads to eat even the thickest of shelled animals.

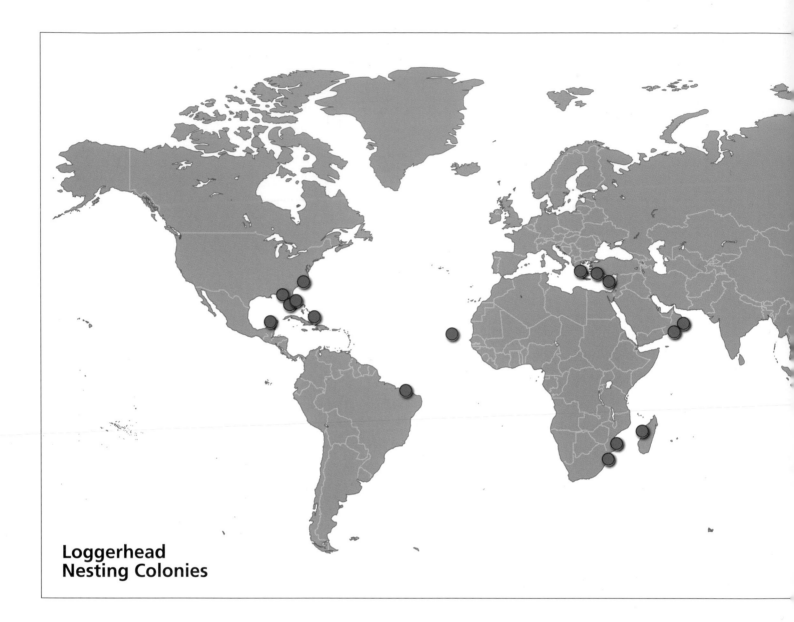

**Loggerhead
Nesting Colonies**

Living organisms such as barnacles, skeleton shrimp, and algae often cover the carapaces of adults and older juveniles, giving them a relatively dull appearance. But upon closer inspection they are full of life. Mobile islands transporting hitchhikers across the globe, they are like floating reefs. Small fish even accompany them on their travels.

Loggerheads are ecologically generalized sea turtles because they have broad ecological requirements and play important roles in ocean ecosystems. They nest over the greatest geographic range of any sea turtle, from the tropics to the temperate zone, and they live and forage in both open waters and in the nearshore regions of all tropical and temperate ocean basins.

Distribution

Loggerheads feed in the Atlantic Ocean from Brazil to Canada, but the greatest concentrations of loggerheads live along the coasts of Mexico, Cuba, the northern Ba-

Population Estimates of Loggerhead Turtle Colonies

Location	No. of Nesting Females per Year
Atlantic Ocean	
United States	
Southern Florida	19,000
Southwestern Florida	1,300
Florida panhandle	155
Northern Florida–North Carolina	1,600
Brazil	1,000
Cape Verde Islands	1,000
Bahamas	130-150
Cuba	75
Caribbean and Gulf of Mexico	60
Mexico	
Quintana Roo and Yucatan	360-580
	Subtotal: 24,680-24,920
Indian Ocean	
Oman	
Masirah Island	12,000
Arabian coast	500
Halaniyat Islands	700
Western Australia	
Dirk Hartog Island	800-1,500
South Muiron Island	150-350
Northwest Cape	50-150
Southern Africa	
Tongaland, South Africa	430
Mozambique	200
Madagascar	100
	Subtotal: 14,930-15,930
Mediterranean Sea	
Greece	1,526
Turkey	683
Cyprus	286
Israel	15
	Subtotal: 2,510
Pacific Ocean	
Japan	700
Eastern Australia	500
	Subtotal: 1,200
	Total: 43,320 44,560

hamas, and along the coast of North America from the Mississippi River along the Gulf coast and up the U.S. east coast to the Canadian Maritime Provinces. Very few loggerheads forage along the European or African coasts. Most Atlantic loggerheads nest in Florida, laying more than 67,000 nests a year—the largest concentration of nesting loggerheads in the world. Nesting also extends north to the Carolinas and occasionally to Virginia. In addition, about 1,200 nests are laid in Mexico, 4,000 in Brazil, and more than 1,000 in the Cape Verde Islands.

In the Mediterranean Sea, loggerheads forage in the eastern and western basins in concentrations between the Balearic Islands and Alboran Sea off Spain and Algeria, along the coasts of Corsica and Sardinia, in the Gulf of Gabès along the coast of Africa, west of Malta, in the Adriatic Sea, off the Nile Delta of Egypt, and in Turkey's Iskenderun Bay. About 45 percent of the juvenile loggerheads in the open waters of the Mediterranean migrate there from the Atlantic. Many of those turtles leave the

CHAPTER ELEVEN

FLATBACKS

Australia's Turtle

THE FLATBACK, *NATATOR DEPRESSUS*, IS A TURTLE OF MYSTERY. WE probably know less about this sea turtle than any of the other six species. It is restricted in distribution primarily to the continental shelf of Australia and its nesting is limited to Australian beaches. We know little about what it eats and no one has tracked its path from nesting beach to foraging ground. In short, we know where they end up but not how they got there. In fact there are only about 100 studies of this species published in the vast scientific literature on sea turtles and many of these are in regional journals of limited distribution. Much of our knowledge on flatbacks is anecdotal and from unpublished reports filed away by government agencies, conservation organizations, and academia.

Famed biologist Samuel Garman first described the flatback turtle as a relative of the green turtle and in 1880 named it *Chelonia depressa*, with *depressa* apparently referring to a shell that looks flat compared to those of other sea turtles. Checking the skeletons on which the original descriptions were based, biologist and taxonomist Peter Pritchard discovered that they actually represented more than one species, and included a green turtle specimen. Perhaps that's why Garman thought the flatback was related and named it *Chelonia*, the genus to which green turtles belong.

As is typical in science, there have been disagreements over the years about how the flatback should be classified. In 1908 Allan R. McCullock thought that he found a new species and gave it the genus name *Natator*, as well as a new species name, *tessellates*. In 1913 D. B. Fry examined a greater range of life history stages and skeletal material. As a result he brought back *Chelonia depressa*. Unfortunately, Fry made mistakes in preparing his hatchling specimens and the broken skulls did look like *Chelonia* skulls.

After that, when flatback skulls were brought into museums scientists saw that they were different and labeled them Kemp's ridleys. The confusion is understandable because they relied on Fry's descriptions and failed to discover his mistake. They did not check the original articles or examine the original specimens used by Garman and McCullock. Shortcuts in science often lead to errors.

By the 1980s Rainer Zangerl and Colin Limpus independently concluded that the flatback was indeed a unique species, that it was more closely related to a log-

Previous pages: Flatback hatchlings stay close to land after leaving the beach. Unlike other sea turtles, flatbacks remain along the continental shelf for their entire lives.

Opposite: A flatback turtle returns to the sea after nesting on Curtis Island, Queensland, Australia. A result of the nesting process, sand sticks to the salty tears below her eye and has accumulated along the edge of the upturned carapace.

(26.8°C). During the nesting season the temperature within the nest increases to more than 95°F (35°C) in the latter part of the nesting season. At Fog Bay sand temperatures can reach more than 93°F (34°C) during most of the nesting season.

Flatbacks nest on exposed, mostly undeveloped beaches that usually have sand dunes, often with vegetation such as trees (Australian pine, *Casuarina equisitifolia*), coastal dune grass (*Spinefix*), convolvulus or morning glory (*Ipomoea*), and low herbs. Generally nesting at night within a few hours of the high tide, flatbacks sometimes nest in the daytime, often in the late afternoon when the high tide occurs near sunset and the beach is cooling off. This is probably to avoid overheating. Their light-colored carapace helps slow down their heating rate. Cloudy and rainy days should pose no overheating problems. No one has studied flatback daylight nesting sufficiently to address these mysteries of temperature regulation.

Flatbacks dig a nest about 20-24 in (50-60 cm) deep at the bottom and about 11-12 inches (28-31 cm) at the top of the eggs. Most nests are placed along the leading edge or at the top of the first dune though some turtles cross over to lay eggs on the secondary dune.

Saltwater crocodiles (*Crocodylus porosus*) eat nesting flatbacks much like those that eat olive ridleys at Playa Nancite, Costa Rica. Crocodiles track down female turtles at night as the flatbacks make their way up the beach to nest. In the morning one can find only the intersecting tracks, some blood, and a few pieces of turtle shell. There has been at least one documented attack on a turtle researcher sleeping atop a sand dune on a flatback nesting beach.

Recently scientists observed a flatback attack on Crab Island, Queensland. A 13-foot crocodile grabbed the turtle in the surf and bit down on it a few times. Then the crocodile rolled over and over until it ripped a flipper off in its mouth, swimming away with its meal. After swallowing the flipper it came back to the edge of the beach, opened its mouth two or three inches, and bared its teeth to the investigators. While it spent the afternoon swimming back and forth along the shore, the scientists stayed on land and kept their distance.

Flatbacks lay very large eggs but fewer of them than other sea turtles. The button-like device on the rear of this carapace is a data logger. It will record depth and time of dives.

From Egg to Hatchling

Incubation takes about 48-66 days depending upon nest temperature. The pivotal temperature for sex determination is 85°F (29.5°C). On most beaches the temperatures at nest depth are 82-90°F (28-32°C) during most of the nesting season. Temperatures within the nest vary by 0.3-2°F (0.2-1°C) during a 24-hour cycle. The heat created by the eggs' metabolism warms the nests by as much as 13°F (7°C) during incubation. This temperature variation should provide a mixed sex ratio among the hatchlings but we don't yet have quantitative studies of sex ratios on nesting beaches.

Flatback embryos can tolerate very high temperatures during incubation. Some of the hottest temperatures ever recorded in sea turtle nests were in flatback nests (98°F/36.5°C), but these temperatures do not reduce hatching success. When the

1.5-oz (42-g) hatchlings emerge from the egg they are almost as large as leatherback hatchlings. They may also carry large amounts of undigested yolk, which offers them an extra food supply until they begin eating in their new ocean environment.

Hatching success is generally high on most isolated island beaches, ranging from 70-95 percent. Larger clutches appear to have a higher hatching success. However, on some beaches predators kill a high percentage of the hatchlings. At Fog Bay a large lizard called the sand goanna or sand monitor (*Varanus panoptes*) destroys 52-67 percent of the nests each year. At Crab Island birds such as rufous night herons

This flatback hatchling shows the distinctive coloration of its species and the large size; compare it to the loggerhead hatchling on page 174.

Tagging Turtles

Sea turtle biologists have been tagging turtles since Tom Harrison started that practice in Malaysia in the 1950s. Sometimes tagging is done with no particular purpose in mind and we say a person has a "tagging reflex." The new investigator or volunteer tags sea turtles because they do not know what else to do. However, in many cases tagging of sea turtles produces extremely valuable information about their biology. For example, we have learned more about the biology of flatback turtles from tagging them on the nesting beach than from any other method.

Colin Limpus, a researcher with the Queensland National Parks System, has possibly tagged more turtles than anyone alive today. His 30 years of effort have yielded some impressive results as well as some cases where tens of

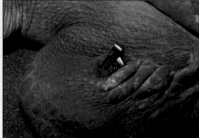

thousands of tags have yielded very little. He actually knows the detailed history of many individual sea turtles because he has tagged them on the nesting beach, captured them on their foraging grounds and recaptured them in subsequent years. He has conducted ultrasounds on the tagged turtles and developed techniques that allow him to predict when they will nest. Studying the same females in the water and on the beach, Colin has taught us much about the reproductive cycles of sea turtles. But it all begins with a little tag that allows him to recognize who's who in Australia.

Top: A Crittercam on a leatherback records the turtle's behavior. Left: Flipper tags are a basic tool used in most sea turtle studies.

(*Nycticorax caledonicus*) kill up to 38 percent of the hatchlings during some periods of the year. These herons cannot swallow the entire hatchling so they peck out the soft body parts and leave the shell. Other avian predators include Australian pelicans (*Pelecanus conspicillatus*), black-necked storks (*Xenorhynchus asiaticus*), white-bellied sea eagles (*Haliaeetus leucogaster*), brahminy kites (*Haliastur indus*), whistling kites (*Haliastur sphenurus*), and ospreys (*Pandion haliaetus*).

Ghost crabs (*Ocypoda cordimana* and *Ocypoda ceratophthalma*) capture only a few hatchlings because the young turtles are too large for the crabs to handle effectively. On the mainland beaches of eastern Queensland, introduced red foxes (*Vulpes vulpes*) used to dig into nests at all stages of incubation and eat hatchlings crossing the beach at night. Dingos (*Canis familiaris dingo*) also raided nests by day and night. Baiting programs, however, have reduced these threats in many areas. On some mainland beaches, however, wild pigs dig up many flatback nests.

We know very little about the early life of flatback turtles. Once they leave the beach, hatchlings disappear into the coastal waters. Instead of swimming off to an

How to Tag a Sea Turtle

The tagging of sea turtles involves several types of simple tags. The first is a clip, either metal or plastic, with a number and notification instructions. The metal clips are more common and typically made of Monel or inconel (both alloys) or titanium. The tag is usually clipped on to one of the front flippers. The Monel and inconel tags have been used for years and have the advantage of preventing the growth of marine organisms on the tag. However, these two tag types seem to mildly irritate the flipper and are often shed over time. Titanium tags remain on turtles more reliably but have the disadvantage that barnacles and other marine animals readily grow on them and can add a great mass of material to the tag. This bio-fouling can also cause the tag to rip out of the flipper.

The plastic flipper tags seem to fall off and break down quite quickly.

A very different type of tag is the passive integrated transponder or PIT tag. It is a small integrated circuit with a tiny antenna that is encapsulated in a sterile glass capsule. The total size of a PIT tag is somewhere in the range of a grain of rice. The PIT tags are easily attached using a needle and syringe injection to the shoulder of a sea turtle. A special battery-powered reader sends a signal to the tag and energizes it. The reader then captures a numerical code from the tag providing the researcher with a unique ID number. The PIT tags were first tried on Kemp's ridley turtles with mixed success. However, researcher John Parmenter demonstrated the utility of PIT tags during a study of flatback turtles at Wild Duck Island. He found that 40% of the Monel tags and 37% of the titanium tags were lost over a five-year period and by eight years 100% of the Monel tags and 40% of the titanium tags were missing. The failure rate of PIT tags was less than 1% over nine years. Donna McDonald and Peter Dutton were inspired by Parmenter's success and employed the method on leatherbacks at St. Croix in the U.S. Virgin Islands. Their results and that of others since suggest the PIT tags have been over 98% reliable. The disadvantage of PIT tags is that one needs not only a reader, but also one that is compatible with the inserted tags. Luckily people tagging turtles are making an effort to standardize, but some government laws restrict bias in favor of a single company.

oceanic stage like other sea turtle species, they stay in shallow coastal waters and feed on small jellyfish and other floating objects. It's surprising that we know so little about a sea turtle that remains so close to the land all of its life.

Migrations

Although flatbacks do not seem to venture beyond the continental shelf, they do migrate great distances after nesting. They also seem to return to particular feeding areas after nesting. Two turtles from Mon Repos were recaptured at feeding areas 134 miles (216 km) and 789 miles (1,270 km) north, respectively. Two Peak Island nesters traveled north 807 miles (1,300 km) to Princess Charlotte Bay along the York Peninsula. Another flatback migrated 640 miles (1,030 km) to Cape Bedford. A fourth swam 217 miles (350 km) north to Newry Island where it was caught in a trawl and released 235 days after nesting. She returned to Peak Island 16 months later and nested again. In contrast, flatbacks that nest in Queensland do not migrate through the deep oceanic waters of the Coral Sea.

Flatbacks, like olive ridleys and loggerheads, spend considerable time floating on the surface of the water in a manner that suggests they are basking in the sun. Sea birds take advantage of the situation and rest on the turtles' backs. At times the turtles appear to be asleep. Perhaps they are warming up to digest their food.

A Remarkable Turtle

One flatback that nested at Mon Repos beach in Queensland during 2003 had a very interesting story to tell. Australian biologist Colin Limpus tagged her (number X8473) in 1977 and reported on her in a scientific article published in 1984. Now, 26 years later, she is still returning to lay eggs on the same beach. She has returned to nest 12 times (about every other year, occasionally skipping two years), generally in a 1-mile (1.5-km) stretch of beach. Turtle X8473 lays 2-4 clutches per season and does not show any signs of slowing down her reproductive output. She is still growing, but very slowly—less than 0.1 in (2 mm) annually.

When Limpus examined her ovaries with his laparoscope in December 1982 he confirmed that she had not laid eggs in 1981, a year he also did not see her at Mon Repos. When she returned in 1987 several recent fractures were seen in her carapace, probably from a collision with a vessel or from being hauled up in a shrimp trawl and thrown on the deck. She still has scars from that incident.

Dr. Limpus has spent the better part of his career with X8473, and although he has learned much about her we still do not know where she lives when she is not nesting or what she does when she is in her home waters. We hope she will survive so that we can continue the long-term studies that are needed to unravel her story and those of her species.

How Many Flatbacks Are There?

Most sea turtle species are in serious trouble because they are being collected on beaches, caught in gill nets and trawls, or hooked on longlines. The flatback is an exception. Because its range is restricted to Australian waters, protecting the flatback presents fewer challenges than those of the other species. No international treaties are needed for its conservation.

The Australian government, to its great credit, protects nesting beaches and turtles at sea. Although in the past many flatbacks were killed in shrimp trawls, since 2000 all trawlers in two of the three largest trawl fisheries in Australian waters have had to use TEDs. There are ongoing discussions about the need for TED use in the third.

Today there are about 10,000 flatback turtles nesting in Australia each year, and some 25,000-30,000 mature females overall. Yet all is not well with the flatback turtle. The Australian government and the Queensland state government have listed flatbacks as "Vulnerable" and the Australian Marine Turtle Recovery Plan considers northern Australian flatback populations to be at significant risk from predation. More work is needed to learn about and protect this unique sea turtle. ⌒

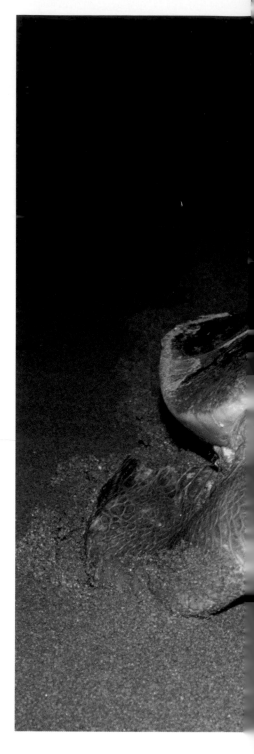

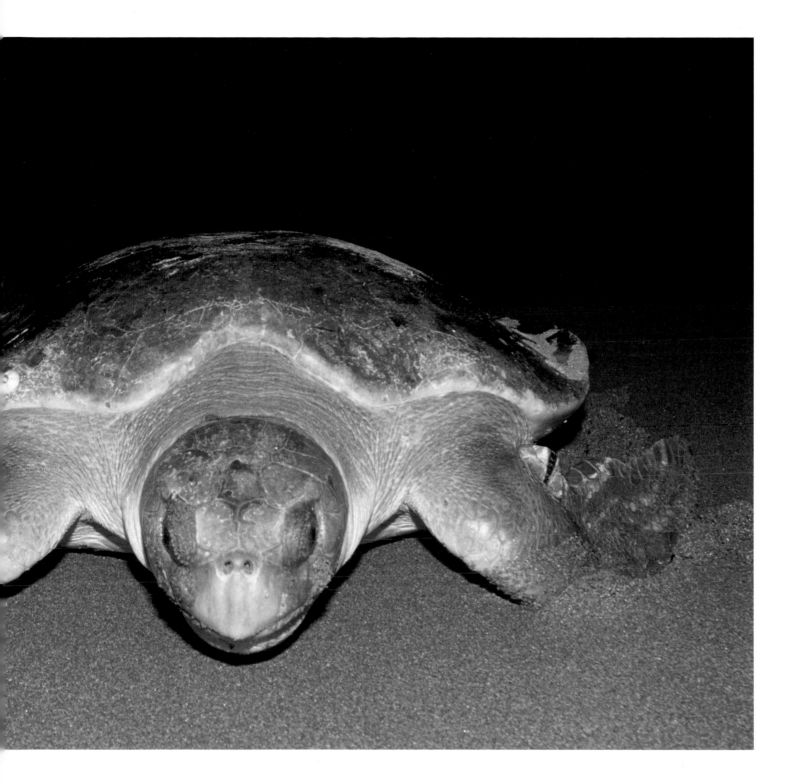

The flatback turtle has relatively soft skin on its carapace which shows the scratches from mating for several weeks. This turtle carries a barnacle on the right edge of its carapace. Even though vigorously protected, this species is still threatened in some areas by trawling.

CHAPTER TWELVE

LEATHERBACKS

The Giant Mariner

THE LEATHERBACK, *DERMOCHELYS CORIACEA*, IS THE MOST EXTRAORDINARY of the shelled marine reptiles. Guillaume Rondelet, a French physician and naturalist, introduced the leatherback turtle to the scientific world in 1554. His book on sea creatures, *Libri de Piscibus Marinis, in quibus verae Piscium effigies expressae sunt* (translation: *Books on Marine Fish, in which True Figures of the Fish are Presented*), put many ocean animals into a more proper scientific context than had been the case, and it remained a standard reference for more than 100 years.

Before Rondelet, leatherbacks were placed in the same category as mythical sea serpents. Rondelet, however, descriptively called the leatherback "*coriaceam,*" which is Latin for "leathery." He described a turtle with "the hide of an ox, tough and black, and which is used to produce bridles, saddles and other leather ornaments." He also called it "Mercury's turtle" because it resembled the shape of the lute that Mercury was thought to have fashioned from the shell of a turtle. Since the time of Rondelet the French have called leatherbacks *tortue luth.*

The leatherback is the largest turtle in the world and one of the largest living reptiles. It is also the deepest diving reptile, able to descend as deep as a whale—4,000 feet (1230 m) and deeper. Fastest of all sea turtles and incredibly maneuverable, the leatherback can swim circles around sharks when it needs to. Its streamlined shell creates very little drag in the water and its large body size and thick, fatty insulation allow it to control its body temperature in a way that makes a mockery of the term "cold-blooded reptile." In fact, it maintains a warm constant body temperature in the frigid North Atlantic. Other distinguishing characteristics of the leatherback include: specializes in eating jellyfish, lays the largest eggs of all the turtles, produces the heaviest clutches of eggs, produces the largest hatchlings, migrates the greatest distances, grows faster than any other reptile, and is among the widest ranging vertebrates on the planet.

Like the big whales, the leatherback swims from near the Arctic to the tropics and as far south as New Zealand and the Southern Ocean. However, it exceeds the distribution of whales by climbing out of the sea to lay its eggs on tropical beaches. The leatherback is a one of a kind, a sea turtle's sea turtle, and arguably one of the most unusual vertebrates in the world.

Previous pages: Leatherback turtle provides a ride to numerous remoras off the coast of Florida. The shape of the leatherback shell makes it hydrodynamic, helping it to swim rapidly. The remoras attach to the shell or ride in the slipstream flowing over it.

Opposite: Leatherback turtle eggs collected by conservationists in Surinam who will incubate them in a hatchery for protection. The large eggs are dimpled when they are laid but will take up water from the ground after being placed into the sand and swell to a perfectly round shape.

Distribution

Leatherbacks are distributed around the globe throughout the tropical and temperate oceans and even edge into subarctic waters. In the Atlantic Ocean they live as far north as Newfoundland and Labrador and visit the waters of Great Britain and Norway. They range as far south as Argentina and the tip of South Africa.

Occasionally these turtles swim from the Indian Ocean to the Atlantic and vice versa. This allows for mating between different populations and may be why there is little genetic difference among leatherbacks around the world. Leatherbacks range throughout the Indian Ocean and the Pacific Ocean, and periodically are reported off Tasmania and New Zealand. One leatherback found in Prince William Sound, Alaska, some 20 years ago is now mounted in a museum in Cordova.

The largest nesting colony is in South America, along the coast of French Guiana. More than 7,000 female leatherbacks laid as many as 50,000 nests there in 1988 and again in 1992. Over the last few years the number of nests has fluctuated and may be declining. It is difficult to assess this population because leatherbacks that nest in French Guiana also nest on the nearby beaches of Suriname, Guyana, and probably in Trinidad, as well. Because Guinean beach sand shifts, leatherbacks change their nesting beaches to follow the areas of shifting sand. These changing sand conditions may account for recent declines in the number of nesting leatherbacks in French Guiana and an increase in Suriname and Trinidad. At the very least we should strongly consider the possibility that leatherbacks form a regional colony, in this case a colony of 13,800 mature female leatherbacks that stretches from Trinidad to French Guiana. The second largest colony is in Gabon, West Africa, where 4,300 females nested in the 1999-2000 nesting season.

The largest colony of Pacific leatherbacks nests in Parque Marino Las Baulas (Las Baulas National Park) in Guanacaste, Costa Rica. Between 59 and 435 leatherbacks nest there each year depending upon the El Niño–La Niña cycle in the Pacific Ocean. Other large colonies visit the beaches of Papua New Guinea, where about 100-200 leatherbacks nest each year. As many as several hundred leatherbacks nest on the Andaman and Nicobar Islands off Thailand. Small colonies of leatherbacks nest in the United States, primarily on St. Croix in the U.S. Virgin Islands and in Puerto Rico and Florida. These populations are increasing in numbers.

Leatherbacks are primarily pelagic (open ocean) animals and migrate great distances from their nesting beaches to their feeding grounds. A number of leatherbacks tagged in French Guiana have been found along the east coast of North America as far north as Newfoundland. New World leatherbacks also migrate to Spain, France, and Morocco. Leatherbacks from St. Croix are known to feed in the north Atlantic and the Gulf of Mexico. One tagged female swam from St. Croix to the resort town of Atlantic City, New Jersey, a trip that spanned some 85 days. The waters of the Grand Banks in the north Atlantic and the Georges Banks off Nova Scotia and

Leatherback turtle returns to the sea after nesting on a beach on Huon Gulf, Papua New Guinea. Leatherbacks heat up when nesting, and the skin along their neck often looks pink due to increased blood flow.

Leatherback Nesting Colonies

Newfoundland are major feeding areas for leatherbacks because of their abundant jellyfish populations during the summer.

My colleagues Ed Standora, Frank Paladino, Steve Morreale, and I once placed satellite transmitters on leatherbacks on the Caribbean coast of Costa Rica and found that they traveled across the Caribbean to the Gulf of Mexico and the Atlantic Ocean. One of those animals made it to the island of St. Kitts where it died in a local fisherman's gill net. Ed later tracked the still-working transmitter to a home on the island. It was hanging on the living room wall, still attached to the rear tip of the carapace, all the while beaming its location to orbiting satellites.

Leatherbacks that nest on the Pacific coast of Mexico and Costa Rica migrate south to feeding grounds off the coasts of Peru and Chile. Our satellite transmitter studies revealed a "biological corridor" running from Costa Rica to Cocos Island and

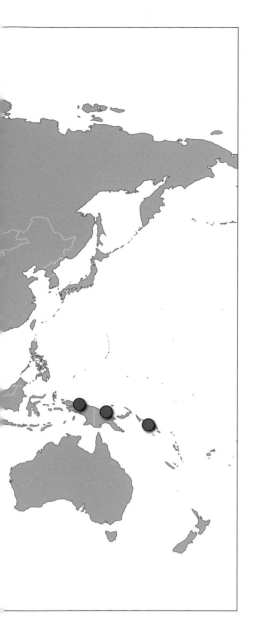

Major Nesting Colonies of Leatherback Turtles around the World Today

Location	Nesting Females per Year	Female Population Size
Western Atlantic		
French Guiana	2,200	
Suriname	1,575	
Trinidad	2,000	
Guyana	200	
Brazil	20	
Subtotal	**5,995**	**13,800**
Eastern Atlantic		
Gabon	4,300	
Subtotal	**4,300**	**9,890**
Caribbean		
Costa Rica		
Tortuguero-Limon	400	
Gandoca/Manzanillo	300	
Panama, Bocas del Toro	200	
Columbia	250	
Dominican Republic	300	
St. Croix, USVI	120	
Isla Culebra	30	
Puerto Rico	30	
Florida, Atlantic coast	40	
Subtotal	**1,670**	**3,840**
Indian Ocean		
South Africa, KwaZulu Natal	120	
Sri Lanka	20	
Andaman and Nicobar Islands	1,000	
Subtotal	**1,140**	**3,420**
East Pacific		
Costa Rica	180	
Mexico	60	
Subtotal	**240**	**910**
West Pacific		
Irian Jaya, Bird's Head Peninsula	650	
Papua New Guinea	300	
Solomon Islands	100	
Malaysia	0-2	
Subtotal	**1,052**	**4,000**
		Total: 35,860

on to the Galapagos Islands. Leatherbacks swam in a straight line, essentially, within the corridor, making the long trip to the Galapagos in 60 days. Biologist Scott Eckert and his colleagues found that turtles equipped with satellite transmitters in Mexico swim southwest and then south to the Galapagos and the waters off South America. Across the Pacific, leatherbacks from Papua New Guinea swim northeast and travel to Monterrey Bay, California, where they feed on abundant jellyfish in the upwelling waters.

Life History

Each nesting season, female leatherbacks lay 1-10 clutches (7 on average) of 50-100 eggs that weigh 2.7 oz (77 g) each. Turtles nesting for the first time lay subsequent clutches 10 days apart, while females that have nested at the beach in previous years return every 9 days. It takes about 1.5-2 hours for a leatherback to complete her nesting process. She will return to nest after 2-7 years, usually every 2-4.

One of the turtle's unique characteristics is its "yolkless egg." Leatherbacks produce 10-40 small egg-like structures that tend to drop at the end of each clutch of eggs. For years called "yolkless eggs," recently biologist Paul Sotherland verified that these false eggs contain neither yolk nor embryo. He calls them "shelled albumin gobs," or SAGs, the result of excess albumin production in the oviduct.

Leatherbacks nest on open beaches that are often exposed to considerable erosion. This is especially true in the Caribbean on St. Croix, along the coast of Costa Rica, and on Suriname and French Guiana. Formerly remote and undeveloped, most nesting beaches are now threatened by humans.

In the Caribbean leatherbacks emerge from March to June to nest throughout the night. Some in French Guiana remain onshore after sunrise. On the Pacific coasts of Mexico and Costa Rica leatherbacks nest between October and March. If the high tide is at night the turtles emerge two hours before the high tide to about four hours after. If the high tides occur within about two hours of dawn and dusk the turtles emerge all through the night.

Mating may take place along the migration route to the nesting beach but we know that it also takes place right off the nesting beach. The ingenious National Geographic explorer Greg Marshall found this out when he put video cameras (he calls them "critter-cams") on the backs of nesting females. Marshall and my post-doctoral assistant Richard Reina recorded the behavior of females after they left the nesting beach. Males chased them, bit them on the head, bit other males on the shell and flippers, and continuously tried to mount the females. They also bit the camera, much to Greg's displeasure.

Hatching success of clutches in undisturbed nests is only about 50 percent, and on some beaches predators such as raccoons, coatis, and dogs destroy many nests. The low hatching success is yet another sea turtle mystery. It cannot be explained by infertility or nest depth (30-39 in/75-100 cm). Neither do we believe the answer is low oxygen or high carbon dioxide levels in the nest because these values are not as extreme as in other sea turtle nests. The presence or absence of SAGs appears to have no effect. Some have suggested that the eggs may be laden with toxins that increase embryo mortality, but definitive studies have yet to be conducted.

Like with other sea turtles, nest temperature determines hatchling sex, with a 50-50 ratio produced at 85°F (29.5°C). A nesting beach's sex ratio varies with the season but is typically biased toward females. At Playa Grande in Las Baulas National Park, eggs laid in October produce males primarily but eggs laid during most of the nesting season produce females, so the beach produces about 95 percent females during a typical year.

The tiny hatchlings weigh only 1.6 oz (45 g) and quickly become bird food if they try to enter the ocean in the daytime. Early in the morning, frigate birds and pelicans patrol the sea and eat any hatchling entering the water as well as many that have not made it far from the beach.

The leatherback uses its large rear flippers to dig a nest on the beach, laying its eggs about 3 feet (1 m) deep in the sand. It covers the eggs gently and finishes by throwing sand over the area.

Surface sand temperature provides the cue that prevents hatchlings from emerging at the wrong time. When hatchlings heat up to 93°F (34°C) they lose the ability to coordinate their movements. Another four degrees renders them immobile. So when hatchlings are climbing up from the nest after hatching and reach the sand near the surface they will be immobilized if the sand is hot. They remain stiff until the sun goes down and the sand cools somewhat. Then they wiggle around and crawl to the surface. This temperature cue ensures that the hatchlings emerge only at night—unless it rains. Of course leatherbacks tend to nest during the dry season so rain usually does not present a problem. When it does, the birds are there and the fate of the hatchlings is sealed.

Once the hatchlings enter the ocean they swim vigorously for at least six nights and days. They can orient in the earth's magnetic field like loggerheads but are usually carried by the prevailing currents to convergence zones away from shore. Small leatherbacks have to eat their weight in jellyfish each day to support their rapid growth rates. Juveniles appear in areas of coastal jellyfish concentrations when their shells are about 43 in (110 cm) long. Thereafter they seem to exhibit the same migration and feeding habits as adults. Leatherbacks appear in large numbers off Florida in February, off southern Virginia in late April, and in the Gulf of Maine along the Canadian coast during the summer. They appear to follow or anticipate the arrival of large schools of jellyfish.

As the fastest growing reptile, leatherbacks reach adult size in 7-13 years. The largest adult females are in South Africa where shell length averages 63 in (160 cm). The smallest adult females are in the Pacific, averaging 57 in (145 cm), whereas in the Atlantic they average 60 in (153 cm).

Leatherbacks forage for food from the surface down to great depths, preying on jellyfish, colonial siphonophores like the Portuguese man-of-war, tunicates, and other soft-bodied animals. In the Mediterranean leatherbacks feed on the deep-water siphonophore *Apolomia uvaria*. In the Atlantic they feed on jellyfish such as *Cyanea capillata arctica*, *Cyanea lamarckii*, *Rhizostoma octopus*, *Rhizostoma pulma*, *Aurelia aurita*, *Pelagia noctiluca*, and *Chrysaora hysoscella*. Leatherbacks also eat the colonial hydroid *Obelia dichotoma*, the jellyfish *Catostylus mosaicus* in Australia, the tunicates *Pyrosoma* (gelatinous colonies that can grow to six feet in length), crabs (*Libinia spinosa*), and amphipods (*Hyperia medusarium* and *Hyperia galba*). Some prey may be consumed accidentally if they live close to jellyfish.

The largely jellyfish diet not only spurs rapid growth but also fuels the turtles' long-distance migrations and high reproductive output. Jellyfish are made up largely of water but they contain some protein, a small amount of fat, iron, and a variety of vitamins and minerals. Still, the water content is so high that they pose an interesting digestive question: How do leatherbacks eat enough jellyfish and efficiently extract the small amount of energy they contain to feed such a large, active reptile?

The answer is that hatchlings and adults have evolved an extraordinarily long esophagus. From the mouth it extends all the way to the rear of the body before looping back up the side to enter the stomach about a quarter of the way from the front of the body. It can expand to hold a tremendous amount of food and is lined with very large, very strong and stiff papillae that point backward. Papillae are also found in other sea turtles but they are much more pronounced in leatherbacks. These cone-shaped papillae are made of cartilage and are sharply pointed.

Leatherbacks swallow jellyfish whole or bite off large pieces with their pointed beaks and sharp jaws. As it swallows the gelatinous mass the leatherback contracts its throat to expel any ingested seawater through its mouth. The papillae prevent the jellyfish from washing out with the water. Then the turtle literally stuffs the food down the esophagus. In this way it can eat huge amounts of food and push it into the stomach as the earlier portions of its meal are being digested.

The leatherback's digestive system is analogous to putting logs into a fireplace. As the first logs burn, new logs are added. Because protein is easy to digest, leatherbacks efficiently convert the jellyfish into leatherback energy, mass, or eggs.

Anatomy and Physiology

The leatherback's skeleton has long fascinated anatomists. Taxonomist, anatomist, and orthopedic surgeon Anders Rhodin spent many years studying the leatherback skeleton and has discovered that the ends of their bones contain thick cartilage filled with blood vessels. This characteristic is unique among all other turtles, the bones of which have thin cartilages containing no blood vessels. Leatherbacks' special cartilage permits their skeletons to grow quickly because the blood vessels flood the developing bone with nutrients.

Another fascinating aspect of the leatherback skeleton is, of course, a softer, leathery shell that is unique among modern sea turtles. We do not know why they evolved this shell but it could be because it compresses at the great depths to which leatherbacks are known to dive. A hard shell might snap. Leatherback shells are composed of numerous small "dermal bones" that lie above the distinct ribs and just beneath the skin surface. The skin itself is a flexible covering that may excrete compounds that keep the shell clean and discourage ocean plants and animals from attaching themselves.

The muscles of leatherbacks exhibit unusual adaptations. Contrary to what happens in most other animal muscles, the metabolic rate of the leatherback's muscle tissue is high and fairly constant over a wide range of temperatures. Therefore leatherbacks feeding in cold waters can maintain their energy level despite low temperatures.

Leatherback hatchlings are covered with tiny scales and are counter-shaded dark above and white below. They row themselves down the beach with their large flippers and swim for days after entering the ocean.

Leatherbacks swimming off Nova Scotia and Newfoundland, Canada, have a body temperature of 60-78°F (15-25°C) when the ocean is as cold as 40°F (5°C). Should we call them warm-blooded given their ability to keep their bodies warmer than the environment? Not exactly, in the sense we usually use the term, because leatherbacks do not have a high metabolic rate like birds and mammals.

Yet adult leatherbacks have unusual adaptations that keep them warm. An adult leatherback's great size, weighing in at 550-2,000 pounds (250-900 kg), ensures that metabolic heat is slow to leave its body. It can also control its body temperature by shuttling between warm and cold water and using its great mass as a buffer, dampening the rate of heating and cooling.

Adult leatherbacks have other advantages, including a thick layer of fat under their shells and fat under the skin of their shoulders and necks. This offers excellent insulation to prevent heat loss. In addition, they a have a unique arrangement of blood vessels in their flippers where arteries and veins lie next to one another. This provides a counter-current blood-flow heat exchanger that helps retain heat.

One would think that these adaptations would pose a problem for leatherbacks when they are swimming in warm water or nesting on a tropical beach. However, another adaptation prevents them from overheating. When leatherbacks get hot they can pump blood from deep within their body past the layer of fat to the soft skin above it. That is why the skin of a leatherback turns pink when it is nesting. The leatherback is dumping heat by pumping lots of blood to its outer surfaces. Conversely, when leatherbacks enter cold water they can constrict their blood vessels to remain insulated. This combination of large body size, thick insulation, and blood-flow control allows leatherbacks to control their body temperature, staying warm in a cold ocean and cool on a hot beach.

Diving Experts

Our "crittercam" videos show that when a leatherback dives down to 90 ft (28 m) the plastron starts to compress inward. In a deep dive the shell will compress even more so that the lungs collapse, forcing air out of the alveoli and into the respiratory passages. This repositioning of air prevents the turtle from getting nitrogen narcosis—"the bends"—from nitrogen that dissolves into the blood stream from the lungs during a deep dive.

Leatherback lung and respiratory features are similar to those of deep-diving mammals. Like them, leatherbacks probably regulate the amount of blood flow to the lungs during diving and inhale just before diving to maximize their supply of oxygen. Thus they may be able to use the oxygen during the initial phases of deep dives, before they reach depths at which the lungs collapse. When a leatherback returns to the surface it quickly exhales and inhales to replenish its oxygen supply. After a long or deep dive a leatherback may take several big breaths in sequence, dipping its head under water between them before submerging for another dive.

Kartik Shanker: A Leader in India's Sea Turtle Community

Kartik Shanker is a Fellow of the Ashoka Trust for Research in Ecology and Environment in Bangalore, India. He has taken the Indian turtle conservation movement to a higher level by the quality of his scientific research and the effectiveness of his community outreach. Kartik began his sea turtle work as a founding member of the Students' Sea Turtle Conservation Network during 1988 in the southern Indian state of Madras. He led volunteer programs at zoos, education programs at schools and a sea turtle hatchery on the beach. Many of those who participated in his efforts have gone on to become wildlife biologists, and in 1990 a group of fishermen in the adjacent state of Kerala started a sea turtle conservation program modeled after his efforts.

Kartik completed his Ph.D. ten years after founding the student conservation network. After complet-

ing his studies he carried out research on the population genetics of olive ridleys on the mainland of India and the ecology of leatherbacks in the Andaman and Nicobar Islands in the Bay of Bengal. From 2000 to 2002 he helped coordinate a national project on sea turtles for the Government. The latter project involved surveys of nesting beaches, training and capacity-building workshops for local workers, and satellite telemetry of olive ridleys.

During 2003 Kartik started to organize a nationwide network of nongovernment organizations to monitor sea turtle populations, raise awareness about sea turtles, and implement conservation programs with local governments and community groups. He also launched a sea turtle newsletter, called *Kachhapa*, which is distributed twice a year to over 1000 subscribers in government organizations, agencies, nongovernment organizations, academic institutions as well as to conservationists and international subscribers. He also produced a series of sea turtle manuals, is completing a book on sea turtles of the Indian subcontinent and regularly turns out popular articles about sea turtles. Kartik's tremendous impact on sea turtle conservation in India cannot be overestimated. He has become a role model for those seeking to transform themselves from student to conservation leader.

Learning to Do Lamaze

Leatherbacks appear to practice Lamaze breathing during egg laying. The pain of childbirth causes human females, like other mammals, to increase their breathing rate and sometimes hyperventilate during labor and delivery. The increased volume of air passing through their lungs flushes out carbon dioxide and makes their blood more alkaline, causing medical complications from a condition known as respiratory alkalosis.

To counteract this tendency women learn Lamaze breathing techniques—rapid, shallow breaths that reduce the volume of air that passes through the lungs. Leatherbacks breathe the same way when delivering their eggs. They reduce the amount of air they take in to 44 percent of resting levels and increase their breathing rate 1.8

Fisherman Ernest Carlisle cools a leatherback caught in a trap net (also called a weir or pound net) off Rhode Island. This type of fishing gear is turtle-friendly because turtles can come up to breathe in the open-top net box. The fisherman was helping with a research study. After being measured and examined, this turtle was released back into the ocean unharmed.

times normal. This helps to maintain normal levels of oxygen and carbon dioxide in their blood and helps maintain a high level of metabolism while they are covering the nest, throwing sand, and crawling back to the sea.

Genetics Tales

Modern genetics has provided important insights into the biology of all sea turtle species, and the leatherback is no exception. Because leatherbacks have declined exponentially in the Pacific Ocean we thought that the low hatchling success could be a fertility problem. One of my students, Barbara Bell, examined developing eggs under the microscope and found that this was not the case. More than 90 percent of the eggs contained cells of developing embryos, indicating that they were indeed fertile. The leatherback's low hatching rate is still a mystery.

Jennifer Crim, another of my students, made an even more amazing discovery. She examined the DNA from blood samples obtained from 1,000 hatchlings from 20 leatherback mothers on Playa Grande, Costa Rica. Every pattern of mating seemed to be evident. At least two and perhaps eight females had eggs fertilized by multiple males. Twelve of the females seemed content to have the entire clutch fathered by one male. Three of the 12 "monogamous" females, however, seemed to have shared the same male, as the DNA pattern for the father indicated it was the same turtle for the trio of clutches.

For many years scientists have been unable to find any genetic differences among leatherbacks around the world. Recently scientist Peter Dutton used the latest genetic techniques to examine both mitochondrial and nuclear microsatellite DNA to determine that leatherbacks that nest in the eastern and western Pacific are genetically distinct populations. This is a critical discovery because it suggests that it is not sufficient for us to save the leatherbacks in New Guinea or some other western colony to save the genetic diversity of the Pacific populations.

Dutton also discovered that leatherbacks from the western Pacific nesting colonies of Papua New Guinea and the Solomon Islands migrate to the north Pacific and west coast of the United States to feed. Leatherbacks from the eastern Pacific nesting colonies tend to migrate south to the waters off South America. This information could help us understand how to manage these populations and emphasizes the importance of international cooperation and local conservation efforts in saving the leatherbacks in the Pacific.

Leatherback Extinction?

Some years ago, my students and I began to wonder how many leatherbacks there were in the world. We counted up all of the records of nesting females and were astonished to learn that they were declining at an alarming rate. In 1980 Peter Pritchard estimated that there were 115,000 nesting female leatherbacks worldwide, and with many colonies still unknown at that time there were probably even more. In 1994 we estimated that there were fewer than 50,000 female leatherbacks left. We raised an alarm but it is hard to attract attention in a world with many environmental crises.

In 1996 our results were published in a prestigious scientific journal. We reported that there were about 34,500 mature female leatherbacks worldwide. Colleagues raised doubts because the numbers were, at best, estimates. I felt fairly certain the grim news was correct but at that time there was a small possibility that we had erred. Sadly, we soon had yet more data supporting the decline. The number of nesting leatherbacks at Las Baulas National Park in Costa Rica varies from year to year but our best estimate is that the number of nesting females has declined from almost 1,400 in 1988 to about 160 in 2003-2004.

The news from other places supports this grave situation. The 90,000 leatherbacks on the Pacific coast of Mexico in 1980 have virtually disappeared. Leatherbacks in India are essentially extinct and in Malaysia they declined from about 10,000 nests in 1956 to just a pair in 2003. However, recent discoveries have added more turtles to the known population in Gabon and the Indian Ocean. In short, today there are about 35,800 adult female leatherbacks in the world, but fewer than 1,000 in the eastern Pacific.

My conclusion is that unless something is done quickly Pacific leatherbacks (like loggerheads) are headed for extinction. Indian Ocean leatherbacks are in grave danger but will likely outlast their Pacific counterparts. In the Atlantic things are better but only by comparison. Some Atlantic nesting colonies seem stable; others are probably in decline, although the data are not conclusive because year-to-year fluctuations make it difficult to detect trends.

The cause of the leatherback's impending extinction across vast parts of its range is quite clear. First, 50 years of harvesting nearly all of the eggs from many beaches have limited the supply of new turtles. Second, killing adults on the beaches of Mexico, Panama, and other places eliminated many nesting females.

Third, and most significant, is the great slaughter taking place in the oceans. In the Pacific in the 1980s, giant drift nets that stretched for hundreds of miles and hundreds of feet deep decimated turtle populations (and caused harm to dolphin and whale populations). Now that most of those nets are gone, leatherbacks still have to evade coastal gill nets, which are smaller but more focused on the turtles' feeding grounds. Chile's swordfish gill net fishery killed thousands of leatherbacks in the 1990s. Nevertheless, the greater threat is longline fishing.

Biologist and demographer Larry Crowder and his students recently compiled data on the number of longline hooks set for swordfish and tuna in the world's oceans. Pelagic longlines stretch as far as a hundred miles and dangle thousands of hooks from each line. Fishing fleets from 40 nations now set an astounding 3.8

million hooks each day—1.4 billion hooks per year. Their numbers are probably underestimates because they did not count unreported fishing in some nations and the many illegal fishing boats.

Scientists who work onboard fishing boats indicate that the probability of a turtle being caught on a longline is small, perhaps seven loggerheads and one leatherback per 1,000 hooks. However, with all the hooks set each year one can extrapolate that about 220,000-250,000 loggerheads and 50,000-60,000 leatherbacks are caught each year. Of the estimated 32,000 juvenile and adult leatherbacks in the Pacific, it is believed that 20,000, or 63 percent, were caught on longlines in the year 2000. That means that

A fishermen harpooned this 1,150 pound (522 kg) leatherback off Montauk Point, Long Island, in 1951. After bringing it to shore in Connecticut the fisherman hung it up for all to see. The skeleton was later put on display in Yale's Peabody Museum, exhibited next to the dinosaurs.

an individual leatherback is likely to be caught on a longline once every two years.

Even if the mortality from longline capture is as low as the 16 percent estimated by fisheries biologists, the death rate from longlines would drive these turtles to extinction over a relatively short period of time. This is seen in our data from Las Baulas, where the annual rate of apparent mortality (never returning to nest again) for nesting females tagged the first time on those beaches is 25-30 percent.

There is hope, however, despite these numbers. Twenty-two years of conservation efforts at Sandy Point on St. Croix and a 40-year campaign on the beaches of KwaZulu-Natal Province in South Africa have succeeded in building up the leatherback populations on those beaches from about 20 to more than 100 animals nesting per year. These remarkable achievements resulted from protecting the integrity of the physical beach, the nests, and the nesting females.

Similar programs in French Guiana and Suriname may be responsible for leatherbacks having expanded their nesting to other areas in the Caribbean and Florida in recent years. They may also be responsible for the survival of the Atlantic leatherback population in the face of heavy fishing pressure. Nevertheless, it is difficult to see how leatherbacks will survive in the Pacific unless something is done quickly about longline fishing and nesting beach protection.

International Protection

Leatherback turtles are considered "critically endangered" internationally by the World Conservation Union (IUCN) and are protected in Appendix I of CITES (Convention on International Trade in Endangered Species) and Appendices I and II of the Convention on Migratory Species (CMS). They are similarly listed by the United States and by many other nations as "protected" in national legislation (for example, the U.S. Endangered Species Act).

The recently ratified Inter-American Convention for the Protection and Conservation of Sea Turtles, which protects all sea turtles, is a boon for leatherbacks. Leatherbacks are also protected by the Memorandum of Understanding on the Conservation and Management of Marine Turtles and their Habitats of the Indian Ocean and South-East Asia (IOSEA), the Memorandum of Understanding on ASEAN Sea Turtle Conservation and Protection, and the Memorandum of Agreement on the Turtle Islands Heritage Protected Area (TIHPA). These measures may help save the populations nesting in Papua New Guinea and Indonesia. The Memorandum of Understanding Concerning Conservation Measures for Marine Turtles of the Atlantic Coast of Africa is beginning to have an effect and may help to save the leatherbacks that nest in Gabon.

The use of turtle excluder devices helps to save green turtles, loggerheads, and ridleys, especially in the United States, Mexico, and South and Central America, but TEDs need to be modified to accommodate the much larger leatherbacks. In addition, the relentless pursuit of fish by the industrial fishing enterprises of Japan, Taiwan, Korea, Spain, and other nations threatens to eliminate the leatherback from the world's oceans. Longline and gill net fishing must be controlled or stopped. We can continue to catch fish without exterminating sea turtles in the process.

Although nesting beaches in places like Las Baulas Park in Costa Rica and on the Pacific coast of Mexico have received protection, these beaches and others are being threatened by development. If these beaches are developed it won't matter if we get the fishing problem under control. There are many beaches in these countries where hotels can be built but only a few on which leatherbacks can nest successfully. In the worst-case scenario, the turtles can tolerate low-density development back away from the beach. That means building structures away from the beach, keeping lights off at night, keeping people off the beach at night except under the leadership of trained guides, and controlling activities on the beach that could disrupt the turtles and their eggs and hatchlings.

The lessons from Malaysia are clear. Developers built hotels and cottages right on the nesting beaches to accommodate as many as 1,000 people a night who came to see the leatherbacks nest. In addition, Malaysians continued to take the eggs. The result was near-extinction for Malaysian leatherbacks.

People can make a difference by assisting in efforts to oppose development on leatherback beaches and by demanding that their governments get industrial fishing under control. This will be a continuing struggle, but one worthy of our best efforts.

The Future

Our goal should be to restore global leatherback populations so that they can fulfill the ecological roles into which they evolved. We may not be able to accomplish this in countries like India and Malaysia during our lifetimes. However, we must work hard to preserve the genetic diversity that still exists in populations that nest on Pacific beaches. This is a reasonable goal because we know that protection in St. Croix and South Africa has made a big difference in the numbers of turtles nesting in those locations. Progress is possible in other locations as well. We either take the steps necessary to save the leatherbacks today or we lose them forever.

This leatherback, entering the surf after nesting, will face a gauntlet of long lines, gill nets, shrimp trawls, and other fishing gear. Its large flippers, streamlined shell, and protective coloration are no defense against modern fishing techniques.

Las Baulas Park, Costa Rica, is the focal point of one of the best-known struggles to save the leatherback turtle from extinction and may provide a capsule view of the future of leatherback conservation. In the past poachers took all of the eggs from the beaches of Playa Grande, Playa Langosta, and Playa Ventanas. Peter Pritchard discovered the beaches there for science and wrote a management plan for the Costa Rican government that provided a basis for protection. Then the president of Costa Rica declared the beaches a national park.

Local expatriate landowners tried to overturn the decree but in 1995 the national Congress passed a law that made the park permanent. Years of struggle led to the formation of a guides association that takes tourists to see the turtles, helps protect the beaches, and promotes economic development among the local people. A series of park directors worked to improve protection and the current director, Rodney Piedra, has not only protected the turtles but built strong ties in the local communities.

For Las Baulas, forming a U.S. non-governmental organization, the Leatherback Trust, played a key role in providing increased resources for protecting and consolidating the park. The Trust raised more than $850,000 in support of conservation, education, and research at Las Baulas. This provided for legal action to enforce national laws, developing a new management plan, and hiring and training rangers and biologists for the park. Frank Paladino provided the intellectual underpinning and the personal strength to focus research and conservation efforts at the park. Scientists, students, and volunteers supported by the Earthwatch Institute came to the park to provide basic protection and to carry out sophisticated research on leatherback biology.

This combination of international support for research, conservation, and capacity building in the park and local community has made Las Baulas National Park a model for leatherback protection. The struggle there continues, however, as developers continue to launch attacks against the park. But over the last 12 years hundreds of volunteers have come to save the leatherback turtle, scientists and students have come from many countries to conduct conservation and research, and local people have been empowered. The turtle is beginning to win.

Despite all of this effort the leatherback turtle is still in imminent danger of extinction in the Pacific Ocean. Las Baulas represents the leatherback's Guadalcanal or Valley Forge. It is either the last stand for the leatherback or the first step back from the road to extinction.

This model can be expanded to other locations. We can take the beaches back one at a time. Friends of the leatherback are working in Mexico, Papua New Guinea, French Guiana, Gabon, Trinidad, and elsewhere. Fishing practices can be changed. Attention can be focused on the leatherback until the glaring light of public opinion forces nations and corporations to change the practices that are driving this species to extinction before our very eyes. The road back will be long and hard but there is still hope that the leatherback, the Giant Mariner, will prevail. ᔕ

A leatherback turtle heads for the sea at Playa Grande, Costa Rica, at dawn after a long night of nesting. The beach is protected as a national park, but real estate developers are constantly trying to change its status.

CHAPTER THIRTEEN

CONCLUSION
Dreaming of Eden

S EA TURTLES ARE MAGNIFICENT ANIMALS, EACH SPECIES FULFILLING unique ecological roles in the world's ocean communities. The "Magnificent Seven Species" are descendants of a small turtle from eastern Brazil, *Santanachelys gaffneyi*, which first ventured into the ocean 110 million years ago. It was a transitional animal that looked like a cross between a modern freshwater turtle and a small sea turtle. It had paddle-like front limbs and movable toes. Its greatest innovation was a large gland behind each eye that removed excess salt it took in from seawater.

Since the appearance of that original ancestor, many different sea turtles have come and gone. The hard-shelled sea turtles arose in the early Cretaceous Period, the leatherbacks in the late Cretaceous about 100 million years ago. Leatherbacks reached their greatest diversity about 50 million years ago when at least six different species lived in the oceans worldwide. The modern leatherback survived probably because it is one of the few animals that eat jellyfish.

For eons sea turtle hatchlings have been crawling up out of the nest, rushing past a gauntlet of predators from the beach to the waves and out to the deep sea. Presumably using magnetic crystals in their brains, the survivors orient themselves within the earth's magnetic field and swim off to spend 10-35 years at sea growing into an adult. Somehow they usually find their way back to the same beach on which they hatched

Swimming rapidly into the future, this hatchling reminds us of what we need to do.

so many years earlier and start the cycle all over again. Under natural conditions only about one in one thousand pulls off the miracle, but that is enough to keep the populations stable. It is a tough life to be a sea turtle, and it's getting harder every day.

When humans came on the scene the oceans were full of sea turtles. More than 600 million green turtles plied the warm shallow seas and 4-5 million hawksbills munched on coral reefs. They shared the ocean with 500 million olive ridleys, 400,000 Kemp's ridleys, and tens of millions of loggerheads. Another 1-5 million leatherbacks swam the oceans devouring jellyfish. The billion or so sea turtles helped make the ocean an aquatic Garden of Eden.

Today we struggle to save the last remnants of this great tribe. It's big news when 15,000 green turtles nest at Tortuguero, Costa Rica. We crow about the 2,500 Kemp's ridleys that return to nest at Rancho Nuevo on the Gulf coast of Mexico. Short-term profits from industrial fishing are balanced against the imminent extinction of the last 4,500 female loggerheads and 4,900 female leatherbacks in the Pacific Ocean. People in the Philippines still eat every turtle that they find nesting on their shores and some Cubans still make trinkets from the shells of the few hawksbills left to catch on local coral reefs.

Yet there are many people—"heroes," I call them—who work on the beaches and in the ocean to study and save these wonderful creatures, God's special gifts. It is difficult and sometimes dangerous work. It is always frustrating and sometimes rewarding. They need your help.

This is the critical century in the history of sea turtles—in fact, the critical century for all of the earth's biodiversity. These beautiful swimmers have survived for 110 million years and yet are threatened with extinction after only a few centuries of exposure to humans. All seven species made it into the twenty-first century, and if we act our grandchildren can look back and say that grandma and grandpa took a stand that got them through it. They will say that we saved the beaches, one beach at a time, that we stopped the slaughter in the oceans, one fishing boat at a time, one country at a time. They will say that we saved sea turtles from extinction. It can happen, and we can do it.

If you are a student and care about the future of sea turtles you should study to be a biologist or conservationist. If you are older, join a volunteer group. Anyone can go on an expedition sponsored by a group such as the Earthwatch Institute and help conserve sea turtles. Write your senators and representatives in Congress. Contribute to sea turtle conservation groups.

Whatever you do, don't accept what might seem "inevitable." Extinction is *not* inevitable. Don't stand by while the last of these beautiful creatures disappears. Extinction is inevitable only if we sit back and watch it happen on the evening news.

We have choices and we make history. What do you want to leave to your grandchildren? Get involved and make a difference. We can leave the world a better place than it was when we found it. Join us on the nesting beaches in either body or spirit. I'll be there, working and waiting. ∾

After many years of conservation efforts led by the Caymen Island turtle farm, green turtles are starting to repopulate the coastal areas around the islands. Their return suggests we may have turned a corner in our relationship with these magnificent animals. One beach, one coastline, one ocean at a time, we can make a difference.

SOURCES AND SUGGESTED READING

GENERAL REFERENCES

Bjorndal, K. A. (ed) 1981. *Biology and Conservation of Sea Turtles.* Smithsonian Institution Press, Washington, D.C. 583 pp.

Carr, A. 1952. *Handbook of Turtles: The Turtles of the United States, Canada, and Baja California.* Cornell University Press, Ithaca, NY. 542 pp.

Carr, A. 1956. *The Windward Road: Adventures of a Naturalist on Remote Caribbean Shores.* Alfred A. Knopf, New York, NY. 258 pp.

Carr, A. 1967. *So Excellent a Fishe: A Natural History of Sea Turtles.* The Natural History Press, Garden City, NY. 248 pp.

Davidson, O. G. 2001. *Fire in the Turtle House: The Green Sea Turtle and the Fate of the Ocean.* Public Affairs, New York, NY. 258 pp.

Ernst, C. H. and R. W. Barbour. 1989. *Turtles of the World.* Smithsonian Institution Press, Washington, D.C. 313 pp.

Lutz, P. L. and J. A. Musick. (eds) 1997. *The Biology of Sea Turtles.* CRC Press, Inc., Boca Raton, FL. 432 pp.

Lutz, P. L., J. A. Musick, and J. Wyneken. (eds) 2003. *The Biology of Sea Turtles, Volume II.* CRC Press, Inc., Boca Raton, FL. 455 pp.

Magnuson, J. J., K. A. Bjorndal, W. D. DuPaul, G. L. Graham, D. W. Owens, C. H. Peterson, P. C. H. Pritchard, J. L. Richardson, G. E. Saul and C. W. West. 1990. *Decline of the Sea Turtles: Causes and Prevention.* National Research Council, National Academy Press, Washington, D.C. 259 pp.

Pilcher, N. and G. Ismail. (eds) 2000. *Sea Turtles of the Indo-Pacific Research Management and Conservation.* ASEAN Academic Press, London, 361 pp.

Pritchard, P. C. H. 1979. *Encyclopedia of Turtles.* T.F.H. Publications, Neptune, NJ. 895pp.

CHAPTER REFERENCES

Introduction: The Seven Swimmers

Meylan, A. B. 1986. Riddle of the ridleys. *Natural History* November: 90-96.

Perrine, D. 2003. *Sea Turtles of the World.* Voyageur Press, Stillwater, MN. 144 pp.

Ripple, J. 1996. *Sea Turtles.* Voyageur Press, Stillwater, MN. 84 pp.

Rudloe, J. 1979. *Time of the Turtle.* Alfred A. Knopf, New York, NY. 273 pp.

Rudloe, J. and A. Rudloe. 1994. Sea turtles: In a race for survival. *National Geographic* 185: 94-121.

Spotila, J. R. 1988. Archie Carr: To the edge of hope 1909-1987. *Herpetologica* 44:128-132.

Life Cycles: From Sand to Sea

Ackerman, R. A. 1997. The nest environment and the embryonic development of sea turtles. In *The Biology of Sea Turtles,* ed. P. L. Lutz and J. A. Musick, 83-106. CRC Press, Inc., Boca Raton, FL.

Bolton, A. B. 1997. Variation in sea turtle life history patterns: neritic vs. oceanic developmental stages. In *The Biology of Sea Turtles, Volume II,* ed. P. L. Lutz, J. A. Musick and J. Wyneken, 243-257. CRC Press, Inc., Boca Raton, FL.

Morreale, S. J., E. A. Standora, J. R. Spotila and F. V. Paladino. 1996. Migration corridor for sea turtles. *Nature* 384: 319-320.

Musick, J. A. and C. J. Limpus. 1997. Habitat utilization and migration in juvenile sea turtles. In *The Biology of Sea Turtles,* ed. P. L. Lutz and J. A. Musick, 137-163. CRC Press, Inc., Boca Raton, FL.

Paladino, F.V. and J. R. Spotila. 1992. Gigantothermy. Dinosaurs and leatherbacks: Standing up to the cold. *Australian Journal of Natural History* 23: 937-944.

Papi, F., P. Luschi, S. Akesson, S. Capogrossi and G. C. Hays. 2000. Open-sea migration of magnetically disturbed sea turtles. *Journal of Experimental Biology* 203: 3435-3443.

Spotila, J. R., E. A. Standora, S. J. Morreale, and G. J. Ruiz. 1987. Temperature-dependent sex determination in the green turtle (*Chelonia mydas*): Effects on the sex ratio on a natural nesting beach. *Herpetologica* 43:74-81.

Standora, E. A. and J. R. Spotila. 1985. Temperature-dependent sex determination in sea turtles. *Copeia* 1985: 711-722.

Van Buskirk, J and L. B. Crowder. 1994. Life-history variation in marine turtles. *Copeia* 1994: 66-81.

Biology: Under the Shell

Bell, B. A., J. R. Spotila, F. V. Paladino and R. D. Reina. 2003. Low reproductive success of leatherback turtles, *Dermochelys coriacea,* is due to high embryonic mortality. *Biological Conservation* 115: 131-138.

Bowen, B. W. and S. A. Karl. 1997. Population genetics, phylogeography, and molecular evolution. In *The Biology of Sea Turtles,* ed. P. L. Lutz and J. A. Musick, 29-50. CRC Press, Inc., Boca Raton, FL.

Carr, A. 1963. *The Reptiles.* Time-Life Books, New York, NY. 192 pp.

Leslie, A. J., D. N. Penick, J. R. Spotila and F. V. Paladino. 1996. Leatherback turtle, *Dermochelys coriacea,* nesting and nest-success at Tortuguero, Costa Rica, in 1990-1991. *Chelonian Conservation and Biology* 2: 159-168.

Lohmann, K. J., B. E. Witherington, C. M. F. Lohmann and M. Salmon. 1997. Orientation, navigation, and natal beach homing in sea turtles. In *The Biology of Sea Turtles,* ed. P. L. Lutz and J. A. Musick, 107-135. CRC Press, Inc., Boca Raton, FL.

Reina, R. D. and P. D. Cooper. 2000. Control of salt gland activity in the hatchling green sea turtle *Chelonia mydas. Journal of Comparative Physiology B* 170: 27-35.

Ridgeway, S. H. (ed) 1972. *Mammals of the Sea: Biology and Medicine.* Charles C Thomas, Springfield, IL. 812 pp.

Schumacher, G-H. 1973. Die kopf- und halsregion der lederschildkröte *Dermochelys*

≈

PHOTOGRAPHY AND ILLUSTRATION CREDITS

Photographs by the following photographers appear on the pages listed: Kyler Abernathy (National Geographic Remote Imaging), 190; Clifton Adams, 69; Bryan and Cherry Alexander (NHPA), 93; Daryl Balfour (NHPA), 171; Fred Bavendam (Minden Pictures), 35; Gary Bell, 24-25, 44-45, 47, 94-95, 103, 162; Annie Griffiths Belt (National Geographic Image Collection), 175; Susana Clusella, 85; Bill Curtsinger, 18, 70-71, 72, 74, 75, 81, 126-127, 133, 134-135, 137, 144-145, 208, 216, 218-219; Bill Curtsinger (National Geographic Image Collection), 82, 86-87, 107, 128, 228; Tui de Roy (Minden Pictures), 143; David Doublet, ii-iii, 12, 49, 110-111, 168-169, 180-181; Nicole Duplaix (National Geographic Image Collection), 105; David Fleetham (SeaPics.com), 40; Jody Johnson, 67, 80, 83, 84, 89, 104, 139, 173, 207; Mattias Klum (National Geographic Image Collection), 91; Tim Laman (National Geographic Image Collection), 108; Frans Lanting (Minden Pictures), 20, 196, 203; Michael Melford (National Geographic Image Collection), 10-11, 174; Anne Meylan, 39; A. Montoya (SeaPics.com), 153; Michael Patrick O'Neill (NHPA), 30-31, 125, 178-179, 194-195; Frank Paladino, 41; Mike Parry (Minden Pictures), 21,199; Doug Perrine (SeaPics.com), 53, 115, 146, 150-151, 158-159, 182, 186, 188, 190, 192-193; Brian Skerry, vi, 160-161, 176, 213; J.R. Spotila, 8, 114, 136; Edward Standora, 1, 214-215; Konrad Wothe (Minden Pictures), 14; Norbert Wu (Norbert Wu Productions), viii, 23, 28, 32, 78, 79, 96, 100-101, 106, 112, 119, 120-121. Illustrations by the following artists appear on the pages listed: David Griffin, 50; Val Kells, 4-5, 6-7, 59, 97, 113, 129, 147, 163, 183, 197; Dawn Witherington, 15, 36-37. International Mapping Associates created all maps. Photographs and paintings courtesy of the following sources appear on the pages listed: ANT Photo Library (NHPA), 189; Department of Library Services, American Museum of Natural History, 56; Chris Luginbuhl, 211; Maria Angela Marcovaldi, 90; Edward Standora, 156; Peabody Museum of Natural History at Yale University, 54-55, 62; Pamela Plotkin, 149; George Zug/National Museum of Natural History, Smithsonian Institution, 60, 61.

An olive ridley turtle returns to the sea.